ATLAS of
GREAT COMETS

Throughout the ages, comets, enigmatic and beautiful wandering objects that appear for weeks or months, have alternately fascinated and terrified humanity. The result of five years of careful research, *Atlas of Great Comets* is a generously illustrated reference on thirty of the greatest comets that have been witnessed and documented since the Middle Ages. Special attention is given to the cultural and scientific impact of each appearance, supported by a wealth of images, from woodcuts, engravings, historical paintings and artifacts, to a showcase of the best astronomical photos and images.

Following the introduction, giving the broad historical context and a modern scientific interpretation, the Great Comets feature in chronological order. For each, there is a contemporary description of its appearance along with its scientific, cultural and historical significance. Whether you are an armchair astronomer or a seasoned comet-chaser, this spectacular reference deserves a place on your shelf.

RONALD STOYAN is Editor-in-Chief of *interstellarum*, and the proprietor of the independent German publisher Oculum-Verlag, which specialises in amateur astronomy books. He is the founding director of the German deep-sky organisation 'Fachgruppe Deep-Sky' and has authored or coauthored twelve books on practical astronomy, including *Atlas of the Messier Objects* and *The Cambridge Photographic Star Atlas*.

STORM DUNLOP is an experienced writer and lecturer on astronomy. He is author of *Collins Night Sky* (2011), a Fellow of the Royal Astronomical Society, and a past president of the British Astronomical Association.

University Printing House, Shaftesbury Road, Cambridge CB2 8BS, UK

Cambridge University Press is part of the University of Cambridge.

It furthers the University's mission by disseminating knowledge in the pursuit of education, learning and research at the highest international levels of excellence.

www.cambridge.org
Information on this title: www.cambridge.org/9781107093492

First published in German by Oculum-Verlag GmbH, Erlangen, 2013.
English edition published 2015.

Printed in Spain by Grafos SA, Arte sobre papel.

A catalogue record for this publication is available from the British Library.

ISBN 978-1-107-09349-2   Hardback

Cover image: Comet Hale-Bopp on 8 March 1997. *Gerald Rhemann*

ATLAS of
GREAT COMETS

Ronald Stoyan

Translated by Storm Dunlop

CAMBRIDGE
UNIVERSITY PRESS

for Renate

Comet Hyakutake 1996. *Kent Wood*

Foreword

Throughout the ages, comets, those extraordinary, wandering celestial objects that, without warning, shine in the heavens for weeks or months, have impressed mankind. The ancient Greeks called them κομήτης, which meant 'with long hair' or 'hair stars'. In other languages they were known as 'broom stars' (in English), or 'tailed stars' ('staartster' in Dutch). The German word 'Schweifstern' ('stars with tails') summarizes their diverse appearance.

The debate about these celestial bodies has been beset with errors and confusion, great ideas and amusing anecdotes. Beliefs and superstitions, art and science have all been influenced by the appearance of great comets. They reflect humanity's development and its search for explanations for appearances in the heavens.

The history of this development is described in this book though the descriptions of the 30 greatest comets from early modern times to the twenty-first century. In choosing these events, astronomical points of view were not the only criteria. The aim was to describe the strange ideas that had the greatest influence on people in former centuries.

This book would not have been possible without the help of numerous individuals. In particular, three people have had a great influence: great thanks are due to Hans Gaab, an expert on the astronomical history of Nuremberg, who placed extensive materials, books and copies at my disposal. His help was particularly valuable in the chapters on antiquity, the Middle Ages and early modern times.

Maik Meyer, who holds the record for the most comets discovered with the SOHO solar probe, and co-author with Gary W. Kronk of the multi-volume *Cometography* series of books, the world's most comprehensive documentation on comets, provided extremely significant help on discovery data and discovery statistics. Burkhard Leitner, the expert on comets on the editorial board of *interstellarum* magazine, provided the charts of the individual comet apparitions.

This illustrated work would not have appeared in such an opulent edition without the wonderful support from numerous astrophotographers. These included Stefan Binnewies, Karl Brandl, Rudolf Dobesberger, Dirk Ewers, Philipp Keller, Michael Kobusch, Bernd Koch, Bernd Liebscher, Jürgen Linder, Norbert Mrozek, Christoph Ries, Gerald Rhemann, Jim Shuder, Wolfgang Sorgenfrey, Peter Stättmayer, David Thomas and Uwe Wohlrab. For their assistance in obtaining picture material, I would also like to thank Prof. Helmut Meusinger (Tautenburg Observatory), Dr Holger Mandel (Heidelberg Observatory), Dr Uwe Reichert and Dr Axel Quetz (*Sterne und Weltraum* magazine) as well as Rainer Mannoff and Wolfgang Sorgenfrey.

I wish to thank Dr Harald Krüger and my editorial colleague Daniel Fischer for their critical checking of the chapter on the modern-day interpretation of comets. Dr Michael Wenzel checked the sections on the history of art. And, by no means last, for their correction of the manuscript I want to thank Dr Winfried Neumann, Susanne Schwab, Angela Hensel and my wife Renate, who has continually sustained me throughout five years of research.

Erlangen, October 2014
Ronald Stoyan

Storm Dunlop, this edition's translator, wishes to acknowledge the help of Chris Caron (on mediaeval German), Françoise Launay (on early French observatories), and Jonathan Shanklin and Steve Edberg (on cometary science).

Table of contents

Foreword 5

Using this book 8

Introduction

Cometary beliefs and fears 10

Comets in art 17

Comets in literature and poetry 26

Comets in science 27

Comet science today 37

Great comets in antiquity 42

Great comets in the Middle Ages 45

Great Comets

Great Comet of 1471 49

Comet Halley 1531 51

Great Comet of 1556 54

Great Comet of 1577 58

Comet Halley 1607 63

Great Comet of 1618 66

Great Comet of 1664 72

Comet Kirch 1680 78

Comet Halley 1682 90

Great Comet of 1744 96

Comet Halley 1759 101

Comet Messier 1769 105

Comet Flaugergues 1811 110

Comet Halley 1835 116

🌠 Great March Comet of 1843 120

🌠 Comet Donati 1858 125

🌠 Comet Tebbutt 1861 132

🌠 Great September Comet of 1882 136

🌠 Great January Comet of 1910 142

🌠 Comet Halley 1910 145

🌠 Comet Arend-Roland 1956 152

🌠 Comet Ikeya-Seki 1965 156

🌠 Comet Bennett 1970 162

🌠 Comet Kohoutek 1973/74 166

🌠 Comet West 1976 170

🌠 Comet Halley 1986 176

🌠 Comet Shoemaker-Levy 9 1994 184

🌠 Comet Hyakutake 1996 188

🌠 Comet Hale-Bopp 1997 196

🌠 Comet McNaught 2007 208

Appendix

Glossary 218

Bibliography and references 219

Index 222

Figure credits 224

Using this book

Text

The apparitions of 30 comets have been chosen for detailed description in this book. With one exception, individual appearances of these celestial bodies are discussed, but Comet Halley is included no fewer than seven times: every one of its apparitions between 1532 and 1986 is individually described.

The text concerning each comet is divided into three sections:

- Orbit and visibility. In this section the apparent path of the comet as seen from Earth is described. The changing position of the comet against the background of the constellations and the conditions under which it was visible to Earth-bound observers are reconstructed. In addition, specific features of its actual path in space are noted.
- Discovery and observation. The circumstances and the date of the discovery of the comet as well as the subsequent observational accounts by various authors are described and summarized. This is where details of the magnitude and length of the tail are to be found, and which, for the most part, rely on estimates by visual observers. Because, in most cases, it is not possible to calculate any objective values, such information may be contradictory. In general, the longer in the past the observations were made, the more critically the information should be viewed. A list of the sources employed is given in the appendices.
- Background and public reaction. This section deals with the scientific findings that relate to each comet, as well as its effect on the general public. Technical terminology is used sparingly, but is, however, not completely avoided. A glossary of the most important concepts is given on page 218.

Data

A table gives the most important astronomical data for each individual comet:

- Designation: The current official designation of the comet
- Old designation: The designation of the comet that was historically employed
- Discovery data: The first sighting of the comet, according to current knowledge (up to 1580 in the Julian calendar, and from 1581 in the Gregorian calendar)
- Discoverer: The first to observe the comet, again according to current knowledge
- Perihelion date: Date of closest approach to the Sun (up to 1580 in the Julian calendar, and from 1581 in the Gregorian calendar)

- Perihelion distance: Distance of closest approach to the Sun in Sun-Earth units (astronomical units, 1 AU = 149 million kilometres)
- Closest approach to Earth: Date of the closest approach to the Earth (up to 1580 in the Julian calendar, and from 1581 in the Gregorian calendar)
- Distance from Earth: Distance of closest approach to Earth in Sun-Earth units (AU)
- Maximum magnitude: The maximum magnitude of the comet according to present-day calculations
- Maximum length of tail: The greatest observed length of the comet's tail
- Longitude of perihelion: The angle between the vernal equinox and the ecliptic longitude of perihelion
- Longitude of the ascending node: The angle between the vernal equinox and the ecliptic longitude of the ascending node of the comet's orbit
- Orbital inclination: The inclination of the comet's orbit relative to the ecliptic
- Eccentricity: The deviation of the comet's orbit from a circular form: $0° = $ circle; $> 0 < 1 = $ ellipse; $1 = $ parabola; $> 1 = $ hyperbola.

Charts

Each appearance of a comet is illustrated with a chart. This shows the course of the object against the sky. The comet's positions and the direction of the tail are shown at fixed intervals. In addition, the beginning of every calendar month is indicated.

The charts also show the position of the comet at its discovery (d), at perihelion (P), closest approach to Earth (E) as well as the final observation (f). The position of the Sun at perihelion is also indicated.

For some comets the whole path across the sky between discovery and last observation cannot be shown. In these cases, the diagrams concentrate on an interval around perihelion and closest approach to the Earth.

Dates are given in the Julian calendar up to 1580, and in the Gregorian from 1581.

Introduction

Cometary beliefs and fears

Ancient ideas

During its changes throughout the year, the starry sky always shows the same set of constellations. The Moon and the planets follow their regular paths. The laws that govern these were discovered early on, so they could then be predicted. Comets, by comparison, arrive and then disappear again. Unlike the planets, they suddenly appear and are unpredictable. They are only ever visible in the sky for a short period of time, whether a few weeks or months, and may move over long distances in a short time. So they are a chaotic, disruptive element in the night sky, which disturbs the orderly procession of the normal movements in the sky.

According to the Roman philosopher Seneca, the Babylonians had already recognised comets as being unique phenomena about 5000 years ago. They saw them as being fiery bodies or planets. Pythagoras, in the sixth century BC, assumed that there was only a single comet, that returned, like a planet, and which also belonged to the planetary sphere and consisted of the mysterious fifth element ('quintessence'). Hippocrates of Chios agreed. According to him, the tail arose through moisture drawn from Earth, which simply reflected sunlight. If the comet appeared in the damp north, then it formed a bright tail, and if in the dry south, by contrast, none. The Greek philosophers Democritus and Anaxagoras held that comets arose by planetary conjunctions. Heraclides Ponticus held, on the other hand, that comets were high clouds reflecting sunlight.

The Greek astrologer, Epigenes, in the second century BC, distinguished two types: the lower, that were created by eddies in the air, and those that arose from vapours. But Apollonius of Myndus believed that comets were planets that came from distant regions of the universe, and were visible over only part of their path. He also assumed that there were very many of them.

Aristotle comprehensively eliminated the many competing and contradictory theories. His conception of comets, which he created about the year 350 BC, was held for 2000 years to be the answer to everything. Aristotle promoted a finite universe. Within the innermost four spheres lay the Earth, surrounded by spheres of water, air and fire. Above these lay the spheres of the Moon, the Sun and the planets. Above everything was the unchanging sphere of the fixed stars.

Comets, according to Aristotle, were part of the lowermost spheres, and thus were not part of the heavens, but were to be considered as weather phenomena ('meteors'). They were created by vapours from the Earth, that rose when warmed by the Sun and were linked with the evaporation of water. If they reached the fourth sphere, they were ignited by friction, and were dragged into a circular path around the Earth. Shooting stars were comets that were particularly easily ignited.

As comets arose from moisture from the Earth, storms and droughts followed their appearance. Aristotle maintained that most comets appeared in the Milky Way, because this would favour terrestrial 'exhalations', whereas the Sun and planets prevented them from forming near the ecliptic. While most authors in antiquity and the Middle Ages followed Aristotle, there were contradictory ideas. In his *Naturales quaestiones* Seneca opposed Aristotle's view. He also rejected Apollonius's planetary nature of comets, because they appeared far from the ecliptic and were not brightest when they were closest to

In all ages and in all cultures, comets engendered anxiety and fear. Deptictions from c.1707–1710. *Anonymous (left), Christoph Weigel (right)*

the Earth. He envisaged comets, however, as long-lasting bodies that were on closed paths, and recommended long-term observations to determine their return.

During the Middle Ages, Aristotle's views persisted, because they were adopted by the Church and were advanced as the expert opinion by influential theologians such as Albertus Magnus (Albert of Cologne). The translation of the Greek text of *Meteorologica* into Latin around 1156 also contributed to this. In this connection, in the thirteenth century, Robert Grosseteste thought that comets were vaporized fire, attracted from the planets as if by a magnet. Around 1350, Conrad of Megenberg held that comets were not just moisture from the Earth, but also arose from people's blood, who would thus be dried out and 'heated'. John of Legnano (Giovanni da Legnano) said that comets made people choleric in nature.

Cometary astrology in antiquity

Among the Babylonians, astronomy was not considered separate from astrology, and this was even more applicable to the interpretation of comets in antiquity. Seneca himself said "If a rare and extraordinarily formed fiery phenomenon becomes visible, then everyone wants to know what it means. Everyone forgets everything else, and asks whether the newcomer should be treated as a marvel or something to be feared. Prophets soon arise, who generally proclaim it as an ominous sign, and the thirst for knowledge and the wish to discover the truth follows them, as to whether it is a sign of foreboding or merely a heavenly body."

The belief was widespread that comets were human souls rising from the Earth to Heaven, but that only those with power and authority shone so brightly that they could be seen from Earth. An example was the comet that appeared after the death of Caesar. Shakespeare drew on this for his famous words "When beggars die, there are no comets seen."

Pliny the Elder, in his *Naturalis Historia*, gave detailed instructions regarding the astrological interpretation of comets. According to his account they generally appeared to the north of the Milky Way. Some comets move. Their significance may be drawn from the constellation in which they lie, the direction of their tails, and their shape. Ten different types of comet may thus be distinguished, which are (in Wittstein's translation):

- Pogonias: bearded stars, because hanging below them is a long beard like those of men
- Acontias: these are called arrow stars because they move swiftly, and their prognostications are swiftly fulfilled
- Xiphias: sword stars. These are the faintest of all stars, and they shine like a sword and throw off small rays
- Disceus: disc stars, which, as the name suggests, are disc-like in shape, have a bright yellow colour and give off very few rays
- Pitheus: have the shape of a barrel, with a smoky light within them
- Ceratis: horn stars, appearing like a horn
- Lampadias: torch stars, resembling a burning torch
- Hippeus: horse stars, horses' manes that circle them extremely rapidly
- Argenteus: white comets, with silvery tails, and so brilliant that it is hardly possible to look at them; which reveal an image of the Divine in a human face
- Hircus: others are rough like wool, and surrounded by a cloud.

The most influential astrological work was undoubtedly the four volumes published by Claudius Ptolemy, known as the *Tetrabiblos*. They formed a complete summary of Greek astrology, supplemented by Ptolemy's own astrological findings. According to this, all appearances in the heavens – and thus not just comets – could be attributed to the planets according to their colour. A dark tint was linked to Saturn, white to Jupiter, red to Mars, yellow to Venus and changes in colour to Mercury. The astrological significance then corresponded to that of the appropriate planet.

Comets were generally associated with Mercury and Mars. As a result they portended wars, hot weather and rebellion. The regions of the Earth that would be affected could

A comet was also viewed as a negative sign by the Aztec emperor Montezuma. He may have predicted his fall in a comet of 1519–1520. *Diego Duran*

Chinese ideas about comets

In China comets also had a great astrological significance, because heavenly phenomena were supposed to presage the country's fate. The Chinese emperor maintained court officials who were both astronomers and astrologers, who constantly observed the heavens and who interpreted the signs as advice for the emperor.

Because the traditional Chinese sky was divided into 283 constellations, which were much smaller than their western counterparts, accurate reconstruction of cometary paths is possible. The positions of comets were additionally given in terms of 28 lunar 'houses', whose boundaries were fixed relative to specific reference stars. Chinese cometary observations are, until early modern times, the most reliable source of information about cometary apparitions. Chinese cometary notes before 213 BC were largely destroyed.

In China, too, the astrological significance of comets was distinguished by their shape. A list of 29 different types of 'broom stars' was given in one source, said to date from 168 BC, but the content of which probably originates in the fourth century AD. Every shape is linked to specific consequences – for example: wars, epidemics, crop failures, famine and revolt. Only two forms are reserved for positive outcomes, which predict the return of the army and a good harvest.

The severity of the consequences was determined by the length of the tail and the comet's lunar house. Chinese astronomers established, by the seventh century at the latest, that comets' tails always pointed away from the Sun,

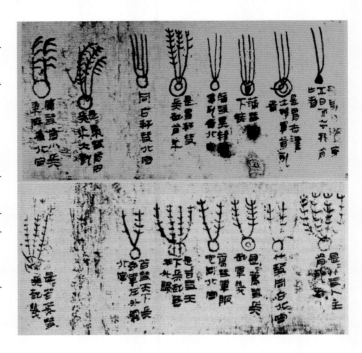

According to Chinese cometary lore, the types of tails enabled conclusions to be drawn about the influence of comets. Most tail forms were given a negative interpretation.

and that comets themselves were non-luminous bodies. The accumulated Chinese knowledge about comets was superior to the European view until the fourteenth century, but always came with an astrological connotation. This has persisted in Chinese popular culture until this day. An allegedly unlucky person – and this characteristic is often ascribed to daughters-in-law who are disliked – is, as a result, described as 'comet'.

be determined from the zodiacal sign in which they occurred, and these signs were indicated by the direction of the tail. In addition, the shape of the head and the tail could predict the type of misfortune. The duration of their appearance, and their closeness to the Sun would, additionally, indicate the beginning and length of the events. Comets appearing in the east therefore led to rapid events, whereas those shining in the west to prolonged ones.

Comets as portents of disaster

In antiquity, in parallel with the astrological interpretation, the idea arose that comets were a sign of negative changes. For one thing, they were seen as prodigies, disturbances of order, that had to be re-estab-

lished by the general public. But a comet could also be an omen affecting a particular person or event. Then intervention was required to avert the disaster.

Even Pliny believed that comets were terrifying signs, 'punishment' that was difficult to avoid: "It is thought that their influence depends on the region they rush across; the strength of which star they absorb; what objects they look like; and where they appear. If they appear like flutes, they are related to music; but to obscene matters when they appear in the private parts of the constellations; to understanding and learning if they form 3- or 4-sided equiangular figures with nearby fixed stars; to the making of poisons when they appear in the head of the northern or southern Serpent."

The Comet Egg

A particularly absurd product of cometary beliefs even involved the Paris Academy of Sciences: In Rome, on 11 November, according to later reports on 2 December 1680, a hen, which had never before laid an egg, had 'with great noise' laid an egg that carried a comet-like design – just at the time when the Great Comet of 1680 stretched across the sky.

Because it was calculated that the 'wonder' had appeared at the seat of the Pope, the Protestant pamphleteers, in particular, saw it as an urgent warning to Catholics to renounce their old-fashioned beliefs. It was thus eagerly taken up and widely depicted. Even in Paris, this grotesque idea was accepted even though it was explained scientifically that it was in no way to be taken as a likeness of the comet – nevertheless people in France widely viewed hens as cometary prophets. The event recurred just two years later, when a hen in Marburg laid a 'comet egg' at the return of Halley's Comet.

A contemporary representation of the 'comet egg' and its source.

The rulers' court astronomers, however, did look for positive indications from comets, because it was not always clear whether comets should be considered part of, or a disturbance of, the divine order. Seneca, in his *Naturales questiones* opposed the interpretation of comets as presages of disaster. He viewed them as part of the divine order, that could not be understood by humans.

In most other cultures around the world, occurrences of comets were interpreted as negative signs. Independently of one another, the celestial appearances were linked to drought, enemies, illness and death.

The Australian aborigines, for example, became worried at the appearance of a comet. It was seen as a bad sign that might portend death, malevolent spirits or evil magic. The New Zealand Maoris called comets 'Auahi-roa' or 'Auahi-turoa', which meant 'long smoke'. Among the Aztecs, too, who called them 'citalinpopoca' ('smoking stars'), comets were taken as a sign of disaster. Allegedly, a comet appeared in 1517 or 1519 before the Spanish invasion. It was, however, initially interpreted as a portent of the fall of the Aztec culture.

The Christian fear of comets

Based on astrological tradition, as well as the belief in portents that were held in antiquity, towards the end of the Middle Ages an interpretation of comets developed that was based on Christianity. Calling on biblical references, comets were interpreted as a sign from God of the imminent end of the world. In the Gospel according to Luke, we have: "... there will be great earthquakes, and famines and plagues in many places; in the sky terrors and great portents." (Luke 21:11) And in the Revelations of John there are suggestions of comets in the description of the end of the world: "... and a great star shot from the sky, flaming like a torch ..." (Revelations, 8:10).

In the sixteenth century, driven by the discovery of printing and the increasing literacy of the general public, the peak of Christian cometary literature occurred. Tidings of new comets in the sky were distributed on broadsheets and pamphlets, most equally accompanied by a scriptural interpretation. These news items fell on fertile soil at a time of turmoil and many changes, shaped by the Reformation and the discovery of America. They also reflected the increasing curiosity about phenomena in the heavens, after the 'Dark' period of the Middle Ages. However, the foundation for the comet literature was still the heliocentric Aristotelian/Ptolemaic System, in which a comet was interpreted as a short-term disturbance of the unchanging heavenly spheres, and as such a special sign.

In German-speaking countries, the broadsheets were, in general, printed in the great evangelical centres such as Nuremberg, Augsburg, Leipzig and Frankfurt. The authors, frequently priests, gained their living partially from the distribution of these writings, which offered excellent possible earnings for little expenditure. In most ca-

A German coin showing the Great Comet of 1618. Obverse (*left*): The threat of a comet (around rim); A sign will appear: Lu(ke) 21 (beneath image). Reverse (*right*): God grant that the Comet star may teach us to better our lives.

Astrological comet types

Pliny had already developed a classification scheme that was employed until well into modern times. According to Girolamo Cardano there were various types, from the appearance of which one could derive the consequences on Earth:

- Vera: This is very long and thin, accompanies the Sun, is terrible and is a mixture of the natures of Saturn and Mercury; it destroys crops and states and kills noblemen and princes.

- Coenaculum: Another is very large, long and broad, and is of the nature of the Moon and signifies general evil.

- Pertica: The third is large and very long (like the first), but not as wide (as the second). Both of these two have their tails directed away from the Sun, but the third has a thick, dense and rounded tail and signifies a shortage of water and infertility. If it is also linked to Saturn, then deaths follow among the general populace and among old people; if with Jupiter, they will be among rulers and high priests; with Mars, great wars and many deaths; with the Sun, famous rulers will be struck; with Venus, great drought and infertility; with Mercury, deaths by mysterious means or secret acts; with the Moon, deaths among the general populace. It is like the nature of Mars with the Sun or Mercury.

- Miles: The fourth has a great brilliance like the Moon and affects the whole Zodiac through which it passes, and it signifies the formation of sects and parties. It has the nature of Venus and also means great drought and infertility and affects the female sex and those of youthful age.

- Asconas: The fifth is unprepossessing, of the colour of the sky, with a long tail and is of the nature of Mercury. It signifies war, the death of outstanding men, serious illnesses and, similarly, treachery and violence, bad times, denunciations and similar ills.

- Aurora: The sixth is red and has a long red tail, but not so great as the fifth (Asconas) and is of the nature of Mars and signifies heat and drought, famine, war, conflagrations, and generally applies in the hot regions.

- Argentum: The seventh is like pure silver, it flashes so much that it may be seen with the eye only with great difficulty. It is of the nature of Jupiter and means changes in reigns and in public life, which are good, but which are linked to great confusion, and its size signifies rich harvest and foodstuffs, mighty winds and a healthy composition of the air.

- Rosa: The ninth is also large and fashioned like a man, and its colour is a mixture of the sixth (Aurora) and the seventh (Argentum). It signifies the death of noblemen and the mighty, changes for the better and is of the nature of the Sun. And if comets appear at a great solar eclipse, then they very particularly signify strange consequences.

ses, the observational report was accompanied by a biblical quotation, followed by an interpretation of the phenomenon. Characteristically, they ended with a call for penance, because these were promoted by evangelical authors, given that Martin Luther saw comets as 'the work of the Devil'. But Catholic sources also fostered anxiety about comets.

The broadsheets catered for the readers' mood in a way similar to modern sensational journalism. So, on the one hand, they fed anxiety over the negative consequences of comets, and on the other, simultaneously, fulfilled the desire for entertainment. In the end, curiosity about the meaning of inexplicable events was also satisfied. According to circumstances, however, a mixture of scriptural and astrological interpretations was also present, if they fitted in with the author's conception of the world.

On the broadsheets, comets were described as 'the scourges of punishment' or 'torches of wrath'. They were a sign from God to Man and required penance and changing one's ways. David Herlitz, one of the most active authors of the broadsheets, wrote in 1619: "So this comet is like an obvious sign that Almighty God has displayed or revealed, in which He shows us the scourge and the threat of his wrath, with which He will strike all those that do not perform a true penance, and whom he will cast into the Fire." Other signs in the sky such as fireballs, halo phenomena and polar aurorae were similarly viewed. Cometary apparitions, however, remained the most significant signs, because they appear to everyone at the same time and are visible for a few days or weeks. Alongside the broadsheets, cometary sermons were delivered and peals of bells were sounded, to assuage God's wrath.

The peak of Christian fears focussed on comets

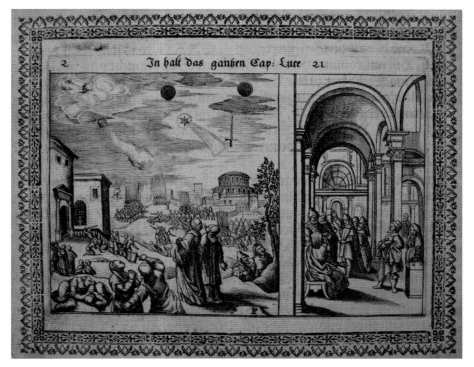

Astrological (*top*) and Christian (*bottom*) themes regarding fear of comets. Whereas astrology saw in a comet itself the cause of a disaster, in the Christian interpretation it was taken with other celestial signs as a warning of the approaching Apocalypse. *S. Lubinietzky (top), Anonymous (bottom)*

in Germany was reached at the time of the Thirty Years' War. Three bright comets were in the sky in the year it began, 1618. They were initially seen as a sign of the defeat of the opposing side, and only after the war were they taken as a portent for the war as a whole.

In the second half of the seventeenth century, comets were strongly regarded as omens of new evils. After 1577 and 1618, the great comets of 1664 and 1680 produced a major response. In this, the troubled times played a large part, such that it was claimed that the danger from the invading Turks was indicated by comets. In the broadsheets, comets were often depicted in the form of a scimitar. In 1680 the Great Comet hung over the besieged city of Vienna. It was alternately taken at a sign of the downfall of the Ottomans or

the menacing danger of war, which could be lifted by penance and amendment. Even extraordinary phenomena such as the appearance of a hen's egg that showed the image of a comet, were seen in the light of a Christian interpretation.

It was not until the eighteenth century that these publications about such marvels came to an end, thanks to the results of astronomical research, which increasingly unlocked the nature of these mysterious comets. By the predicted return of Halley's Comet in 1759, they had largely died out.

Early modern astrological interpretations

Cometary astrology, based on Ptolemy's *Tetrabiblos* reached a peak in early modern times. The view was widespread that comets signified a change, which was dependent on the constellations in which they appeared and through which they passed. The interpretations were thus similar to those concerning the planets. Classification depended on their colour or the star near which they were discovered. Red comets were particularly popular, which were identified with Mars, or comets with dark, yellowish tails, which were identified with Saturn. Individual cometary phenomena had particular influence on countries that were ruled by the corresponding zodiacal sign.

Factors of astrological significance:

- *magnitudo*: the brighter a comet, the greater its effects
- *color*: an indication of the corresponding planet
- *splendor*: the greater its brightness, the greater its effects
- *forma*: the shape of the comet according to Pliny
- *diuturnitas*: the duration of visibility would be crucial for the duration of its effects, consequently various conversion factors were given
- *locus*: the constellation in which the comet appeared. Earth signs led to drought; water signs to floods; air signs signified revolts; and fire signs war
- *motus*: the direction and velocity of the comet indicated the regions affected
- *habitus ad solem*: the position of the comet relative to the Sun indicated the start of its effects: these arrived quickly with comets to the east of the Sun, and slowly for those west of the Sun
- *situs orbi*: the projection of the comet's path onto the Earth indicated the regions affected

Kepler, who alongside his pioneering astronomical discoveries, also practised as an astrologer, summarized the multitude of factors involved: "standstill or velocity; location of the tail; colour; brightness or darkness; an orderly or irregular motion; how and also from which heavenly house or sign it emerges; whither it is headed; where it vanishes; which constellations were at its beginning and end; how long it persisted; the locations over which the head passed, and which loca-tions or birth horizons coincide with its path, together with all other such things." Kepler did not see comets as completely deterministic, but rather as a disruption of heavenly harmony. Comets were sent by God. Generally speaking, it could be held that 'Comets, wars, pestilence, price rises, earthquakes, droughts (or on the other hand, floods) commonly arrived together.'

The astrological view contradicted and was openly in conflict with Christian teaching, in which God alone could foreordain fate, and not phenomena in the heavens, and also that a man's actions had an influence on his fate. Nowadays, comets have hardly any significance in astrology. The German astrological portal astrowiki.de states: "The significance of comets as a fascinating heavenly phenomenon is probably greater than their astrological relevance."

Many astrologers, however (even today), view comets as a sign of changes on the personal and political level. Comet Kohoutek in 1974 was linked to the coup d'etat in Chile, for example, as indicated by its discovery in the constellation of Sagittarius. Its influence was also seen as negative for Portugal, because the 'national sign' of Pisces, together with the location of the comet, determined the loss of the colonies. In 1986, Comet Halley was linked to the Chernobyl disaster. With Hale-Bopp in 1997, astrologers drew parallels with Dolly, the cloned sheep.

Modern conspiracy theories

Even after the decline of astrological interpretations and of the belief in marvels, there are still rather daring claims made about comets. In 1773, for example, there was a mass panic in Paris, after the rumour went around that the astronomer Jérôme Lalande had predicted the impact of a comet in a talk that he gave. In 1910, there was panic and suicides when the Earth passed through the tail of Comet Halley, and the cyanide compounds that had been discovered in it shortly before were thought to be likely to kill people. And, sadly, it is still fresh in our memory with the mass suicide of members of the Heaven's Gate sect, who interpreted Comet Hale-Bopp as the divine vehicle to carry them to the beyond, after American conspiracy fanatics maintained that the comet was being followed by a Saturn-like spaceship, shown on photographs – however, this was simply a diffraction effect caused by the optics that were used.

Comets in art

In contrast to the popular perception by which the appearances of bright comets were particularly significantly regarded in all ages, when it comes to comets as a motif in the visual arts, they appear only occasionally. This apparent contradiction is based on the clear restriction of their role as harbingers of disaster and portents, by which comets were regarded until early modern times. Their significance as negative signs was so strong that they could not be represented in any other context.

Back in antiquity, comets were not shown in art, being firmly linked, in iconological terms, to astrology. The few representations of comets were stylized images on coins that were struck at the death of Julius Caesar. Other rulers also had coins struck with comets as a motif.

The Middle Ages

In the Middle Ages, comets, because of their negative connotations, were excluded from art. In particular, they could not be employed to symbolize the Star of Bethlehem. In many sources the few representations of the Star of Bethlehem as a comet were condemned, despite it not being possible to substantiate this assertion. Among these representations is one on an ivory panel on the throne of Max-

imian, Bishop of Ravenna, dated to the years 540–545 AD. There, a very stylized representation of the Star may be seen above the scene of Christ's birth. It looks more like a flower than a comet. Because the 'tail' is also an apparent ornament on other panels, and the connotation associated with a cometary subject contradicts the intended significance, we can be fairly certain that a comet is not intentionally depicted here.

A similar case applies to the set of mosaics by Pietro Cavallini (1250–1330) in the church of Santa Maria at Trastevere (in Rome), which have been dated to 1291. Here the Star is depicted with three rays pointing towards the Christ-child. This is probably intended as a symbol of the Trinity. Because rays around stars in paintings from the Middle Ages should not be taken as physical features, but instead have an iconographical significance, again this should not be seen as the image of a comet.

The famous depiction of Halley's Comet on the Bayeux Tapestry, which has been dated to 1291, should be seen in a quite different light. The Tapestry, 70 m long and 50 cm wide, illustrates events surrounding the Battle of Hastings in 1066, when the Norman, William the Conqueror, wrested the English throne from the Anglo-Saxon

◄ A Roman denarius with the image of the Emperor Augustus and a depiction of the comet of 44 BC, symbolic of the divinity of the Julian family.

▼ The Bayeux Tapestry (detail) with the representation of Halley's Comet of 1066 (*left*). In contrast to this the star shown on the Maximian Throne in Ravenna is nothing like a comet (*right*).

king, Harold. A highly stylized comet is depicted on the join between panels 32 and 33, representing Harold's downfall. The inscription above reads "Isti mirant stella" ('they marvel at the star'). The representation of the comet is notable, because there is a gap between the head and the tail. Is this showing one of Comet Halley's typical breaks in the tail? (See, for example, the photos on page 182.)

The first true, not stylized, representation of a comet is found in the famous fresco Adoration of the Magi by Giotto di Bondone (1266–1337), which forms part of the decoration of the Arena Chapel (Cappella degli Scrovegni) in Padua. It has been dated to 1303–1306. Giotto is rightly regarded as the person who revived the art of painting. For the first time he showed the Star of Bethlehem as a comet. The comet is shown with naturalistic features, which may be ascribed to Giotto's own observation of Halley's Comet in 1301. This representation contradicted the biblical interpretation as a 'good star', because comets were seen as bad omens. A copy of this scene, also with the depiction of a comet, is found in the church of St Francis in Assisi. The scene of the Nativity (or the Adoration) was reproduced here in 1315.

Early modern times

One of the few additional works of religious art that depicts a link with comets is the oil painting The Birth of Christ by Hans Baldung (a.k.a. Hans Baldung Grien, 1484–1545). This shows a stylized small comet above the thatched roof of the Nativity's stable. The tail points towards the Christ-child's head. This representation definitely originated as a result of the influence of the Great Comet of 1471. The oil painting The Flight from Egypt in the Schottenkirche in Vienna, by some unknown master, dates from the same period. This depicts another naturalistic representation, which is probably also related to the Great Comet of 1471.

This is in contrast to the ordinary artistic depiction of comets. One of the most famous works is Melencolia I of 1514 by Albrecht Dürer (1484–1545). Dürer adhered closely to the astrological interpretation current in his time. According to this, melancholy was a property of Saturn. The comet in the background is exemplary in showing its connection to that planet. It is thus rather pale and hangs above the sea, which symbolizes the water sign ruled over by Saturn. Its location under the rainbow, however, refers to Christianity's superiority to astrology. In addition, Dürer depicted the comet in a stylized manner.

At about this period between 1507 and 1513, tempera miniatures appeared in the *Lucerne Chronicle*, painted by Diebold Schilling (c. 1460—1515). These harked back to the *Nuremberg Chronicle*, which depicted the comet of 1471 (although in popular form), and showed four differing images with comets. They depicted cometary phenomena for 1400 (the image shows the crusade against the Turks in 1394 as well as the outbreak of the plague); for 1456 (the earthquake at Naples is shown and a rain of blood over Rome, with the birth of monstrosities being additionally shown in the foreground); for 1471 (the theme here is drought and war, with the depiction of a funeral procession); and for 1506 (floods, landslides and ruined harvests are shown). The depiction of the comets is stylized, but at the same time very imaginative. The tails always point down towards the ground, to emphasize the link with the Earthly events below. All link comets with negative events. In Schilling's drawings we can see the rendering of popular beliefs and a precursor of the later comet broadsheets.

These arose with the introduction of printing towards the end of the fifteenth century. The appearances of bright comets were turned into a sort of sensational literature, with astrological or Christian significance. On them, comets were more-or-less stylized or distorted as a sword or 'scourge of God', to convey this message. Comets also gained popularity as a typical theme for depictions of the Apocalypse and in the iconology of vanity. As a result, they are hardly ever found in any other setting in art from the fifteenth to the eight-

Halley's Comet in 1456, a tempura miniature illustration from the *Lucerne Chronicle*, 1507–13. *Diebold Schilling*

▶ The Adoration of the Magi, a fresco in the Arena Chapel in Padua, 1303–06. *Giotto di Bondone*

The Comet of 1680 over Rotterdam. Oil painting, 1681. *Lieve Verschuier*

eenth centuries, and were, in addition, not suitable for the subject of the Star of Bethlehem in the religious context. These broadsheets persisted until the middle of the eighteenth century. From this period on, the accuracy of the depictions of cometary form increased and, simultaneously, stylization decreased.

The Enlightenment and Romantic periods

From the second half of the seventeenth century, in parallel with the scientific discoveries, the topic of comets broke free from its astrological significance. In the oil painting of the Great Comet of 1680 over Rotterdam by Lieve Verschuier (1634–1686), there is no longer any symbolic content in the foreground, but rather an impression of the celestial phenomenon itself. Curiosity about the comet predominates rather than fear. The observers in the foreground are holding cross-staffs, which were used for the determination of astronomical positions – a symbolic indication of comets as celestial bodies subject to scientific examination. Verschuier's painting is not an exact reproduction, however, but a form of summary of features that the comet showed at various times. They were subsequently reproduced in the painting.

In an oil painting, produced in 1711, by Donato Creti (1671–1749), a remnant of the fear of comets is seen in the demeanour of the woman in the foreground. Otherwise there is rather a relaxed feeling to the painting. This work cannot be ascribed to any particular comet.

From the eighteenth century, the treatment of comets in caricatures is observed. Commercial artists used this form of image, above all, to make fun of political events. Such images were sold in some shops, and at festivities and meetings. Comets thus served simply as a symbol for a political context, in particular in connection with Napoleon, where the classical meaning of comets was employed ironically, with the comet predicting his downfall. The Great Comet of 1811, however, served as a motif of its own. Some caricaturists used it to poke fun at comet hysteria. The publication of caricatures continued for the comets of 1842 and 1858, but at the same time, with the appearance of widespread copying, they assumed a rather romanticized form. The

Comet. Oil painting, c. 1711. *Donato Creti*

▲ Melencolia I. Engraving, 1514. *Albrecht Dürer*

▲ Donati's Comet on 5 October 1858. Watercolour, 1858. *William Turner of Oxford*

◀ Donati's Comet. Watercolour, 1858. *Unknown Artist*

▲ Pegwell Bay, 5 October 1858. Oil painting, 1858.
William Dyce

▶ Donati's Comet over Dartmoor (detail). Watercolour, 1859. *Samuel Palmer*

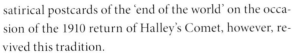

LOOKING AT THE COMET TILL YOU GET A CRIEK IN THE NECK.

satirical postcards of the 'end of the world' on the occasion of the 1910 return of Halley's Comet, however, revived this tradition.

Donati's Comet of 1858, which must be numbered amongst the most beautiful comets in history, and which was visible for a long time, was adopted as a favourite motif by the Romantic movement. In the watercolour by William Turner of Oxford (1775–1851) the comet is in the very centre of the depiction, and the foreground serves merely as a frame for the phenomenon. In the watercolour of the comet by Samuel Palmer (1805–1881), which was painted in 1859, the comet is depicted in a particularly romantic setting over the hills of Dartmoor. Palmer prepared the sketch for his painting on 5 October 1858. The position of the star Arcturus is, in contrast to the rather stylistic sky shown by William Turner, very accurately reproduced.

William Dyce (1806–1864) set Donati's Comet over Pegwell Bay in southeast England in his oil painting. The almost photographic reproduction in the Pre-Raphaelite style, shows the link between the foreground (stooping, individual women) in contrast to the massive background (weathered cliffs) as a metaphor for life's perils. The pale comet above the scene serves as a symbol of the extrater-

Three typical caricatures from the 'comet-mad' nineteenth century. *Anonymous (top left), Honoré Daumier (top right), Thomas Rowlandson (bottom)*

restrial forces in the sky that drive fate that cannot be influenced by human beings.

Modern times

From the beginning of the twentieth century, representations of comets increasingly diverged from one another. The less the motif of comets was seen in astrological or romantic terms, the more freely images could be employed. Painting thus became ever more individualistic.

A wave of comet pictures comparable with that caused by Donati's Comet in 1858 was unleashed by Halley's Comet in 1910. This fell on fertile ground with the early Expressionists. In his painting Comet, Vassily Kandinsky (1866–1944) showed a bright yellow tail over seemingly Christmas scenery, whereas the Armenian Martiros Sarjan (1880–1972) created something reminiscent of an oriental/magical landscape, in which the motif also appears in a reflection. The Ukrainian painter Georgii Narbut (1886–1920), in contrast, depicted a cold Art Nouveau landscape, in which the bright comet appears almost threatening, whereas with Walter Ophey (1881–1930) the theme dissolves entirely into individual points of light.

▲ Comet. Watercolour and gold bronze powder, 1900. *Vassily Kandinsky*

► Comet. Oil painting, 1910. *Walter Ophey*

▼ Landscape with comet. 1910. *Georgii Narbut*

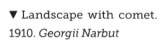

▼ Comet. 1907. *Martiros Sarjan*

Comets in literature and poetry

Just as in art, for a long time the comet motif in literature was bound up with negative astrological meaning. Only with the Enlightenment did it become possible to employ comets in a metaphorical sense in other contexts. Since the second half of the nineteenth century, comets have been a central motif in science fiction.

Early modern times

"When beggars die there are no comets seen / the heavens themselves blaze forth the death of princes." This famous verse from the second act of Shakespeare's tragedy 'Julius Caesar' dates from 1599. Shakespeare compressed the entire literary role of comets into one sentence, which he put into the mouth of Calpurnia, Caesar's wife. The statement is quite in accord with the ancient astrological interpretation.

Until well into the eighteenth century, it was no different in German-speaking countries. Among many similar tales, the 'Kometspiegel' ('Comet mirror') of Thomas Hartmann, written in Halle in 1606, has become famous, and which, in a mixture of Christian fear of comets and astrological interpretation, lists the threats to mankind posed by a comet:

"All comets indeed give evidence / of a lot of bad luck, affliction, peril and danger / and a comet never appears / to be cares without evil meaning / in general eight kinds of affliction occur / when a comet burns in the air:

1) Much fever, illness, pestilence and death,
2) Difficult times, shortages and famine,
3) Great heat, droughts and infertility,
4) War, rapine, fires, murder, riots, envy, hatred and strife,
5) Frost, cold, storms, weather and lack of water,
6) Great increase in people going into decline and death,
7) Conflagrations and earthquakes in many places,
8) Great changes in government.

But for us to do penance from the heart, God afflicts us with disasters and pains."

The Enlightenment and Romanticism

About the middle of the eighteenth century, in parallel with the scientific revelations about comets, the literary point of view also changed. Authors during the Enlightenment discovered comets as an ideal motif. First, however, they had to be freed from the astrological restrictions. In 1742, the French philosopher Pierre Louis Moreau de Maupertuis wrote that "these stars, after having been the world's terror for a long time, have suddenly become discredited, so that they are able to cause no more than a cold in the head."

August Wilhelm Iffand (1759–1814), in his farce 'The Comet' likewise dealt with the overwhelming fear of comets. In a comical fashion this recounts how a charlatan, who as a result of the supposed end of the world threatened by a comet, attempts to gain material and personal advantages, until his fraud is finally exposed in a humiliating and hilarious scene.

Comet Flaugergues of 1811 was immortalized in the famous novel War and Peace by Leo Tolstoy (1828–1910). In it, Pierre, one of the two main characters, sees it "Above the dirty, ill-lit streets, above the black roofs, stretched the dark starry sky. ... Almost in the centre of it, above the Prechistenka Boulevard, surrounded and sprinkled on all sides by stars but distinguished from them all by its nearness to the earth, its white light, and its long uplifted tail, shone the enormous and brilliant comet of 1812 – the comet which was said to portend all kinds of woes and the end of the world." To Pierre, however, the sight, 'invigorated and strengthened' him: "joyfully, he gazed at this bright comet." However, the Russian author got the year wrong, because by 1812 Comet Flaugergues was no longer visible to the naked eye!

Science fiction

Edgar Allan Poe (1809–1849) did not live long enough to experience the brightest comets of the nineteenth century. In the *Conversation of Eiros and Charmion* (1839), the American author was inspired by the end-of-the-world mood of the Adventists in the USA in the 1830s. In it, two dead people discuss the end of the world, which has been caused by a comet that has poisoned the Earth's atmosphere.

Jules Verne (1828–1905) in the story *Hector Servadac* (published in English as *Off on a Comet*) used a comet as a vehicle for his protagonists. The tail, which brushed the Earth, carries various people away with it, and carries them on a tour of the Solar System. Jules Verne's comet 'Gallia' consists, remarkably, of metals – a large fraction being gold – and has active volcanoes. The space-tourists land safely, using two air-balloons, when the comet returns to Earth.

In the novel *In the Days of the Comet* by H.G. Wells (1866–1946), the tail of a comet causes humanity to be 'exalted' into a utopian, peaceful community without property. The author's predictions are amazing; he both foreshadows the forthcoming First World War, as well as the passage of the Earth through the tail of Halley's Comet in 1910. A similar theme is adopted by the German author Hannes Stein, in his German-language book of 2013, *Der Komet*, he describes a world without the catastrophes of the twentieth century.

Nowadays, apocalyptic scenarios involving the impact of a comet are popular. This is reflected by the numerous films on the topic such as the 1998 Hollywood blockbusters *Deep Impact* and *Armageddon*.

Comets in science

Fifteenth and sixteenth centuries

For centuries the nature of comets was not a matter for consideration. Astrological theory was too strongly entrenched. It was only at the beginning of modern times that the fundamental question came to be posed: "Are comets meteorological or astronomical phenomena?" "Are they sublunar, being situated under the Moon, or are they supralunar, lying beyond it?"

These questions should have been answered by the determination of the distance of comets. The principle of distance determination rests on parallax, that is, on the apparent spatial shift in the positions of comets in front of a more distant background (the starry sky), based on viewing the comet from different angles. If comets were closer than the Moon, as Aristotelian doctrine held, then when comets were viewed from two different locations they should appear at different positions on the sky. Similarly, parallax should have been detected with observations from a single, fixed position over a period of some hours, if the comet were closer to the observer than the whole firmament, which was rotating around the Earth.

Before the discovery of the telescope in 1610, people tried to determine the positions of comets with the naked eye. In the Middle Ages, the instruments available were the torquetum and the cross-staff (or Jacob's staff). In early modern times, quadrants were also used, and the larger they were, the more accurate their determinations of positions. Astrolabes and armillary spheres were also employed.

The Italian geographer Paolo Toscanelli (1397–1482), famous for his (inaccurate) chart of the Atlantic Ocean that led Columbus to believe that a shorter route to India lay to the west, was one of the first people to determine the positions of comets. He followed six comets between 1433 and 1472 and recorded their motions across the sky. His notes were not rediscovered until the nineteenth century.

The Austrian astronomer Georg von Peuerbach (1423–1461), was the first to undertake an attempt to determine parallax with Comet Halley in 1456. Because of the shift in the position of the comet relative to the stars over a few weeks, he came to the conclusion that the parallax was less than that of the Moon, and thus that the comet was more distant than the Moon. The method used by Peuerbach was very inaccurate, because the proper motion of the comet itself was not taken into account. The actual parallax of most comets was not capable of being determined with contemporary methods.

The mathematician and astronomer Johannes Müller, known as Regiomontanus (1436–1476), born in Königsberg in Franconia (now Bavaria), a pupil of Peuerbach's, refined the method of determining the parallax of comets. However, he also neglected the comets' proper motion and was therefore only able to give imprecise details, which left open the question of the sub- or supralunar location.

In 1531, Peter Bienewitz, known as Apian (1495–1552), found, following position determinations of Halley's Comet and subsequent comets, that cometary tails are always directed away from the Sun. Apian was famous for the *Astronomicum Caesareum*, a monumental work on astronomy that he published and printed himself.

The Italian astronomer and astrologer Girolamo Cardano (1501–1576) compared the rate at which Comet Halley 1531 and another comet of 1532 moved, with that of the Moon and found them less. Consequently, he located the comets in the supralunar sphere. Cardano believed that comets were in the shape of a lens and were illuminated by the Sun, with the tail being formed by the optical effect of the lens. Cardano is also known for his horoscope of Jesus, for which he was taken into custody by the Inquisition in 1570.

The Danish astronomer Tycho Brahe (1546–1601), famous for his accurate determinations of position, followed the Great Comet of 1577 and obtained the longest and most accurate series of positions secured up to that time. He succeeded in determining the parallax of the comet over a period of three hours, as did Thaddaeus Hagecius (Tadeáš Hájek, 1525–1600) in Prague and Cornelius Gemma (1535–1578) in Leuven. Tycho obtained an angle of 15' as a maximum, and concluded that the comet had to be at least 230 Earth radii distant. He also rejected Copernican theory and believed in a mixed geocentric/ heliocentric universe. He accepted an oval heliocentric orbit for the comet. Like many of his contemporaries, however, he also believed in its astrological significance.

The Schwabian theologian and mathematician Michael Mästlin (1550–1631), a student of Apian's son, Philipp (1531–1589), also followed the Great Comet of 1577 and was unable to determine any perceptible parallax. As a result he viewed it as being a celestial object and calculated a circular heliocentric orbit. So at the end of the sixteenth century, the view became established that comets, like the planets, lay beyond the Moon.

The seventeenth century

Previously, straight-line or circular orbits were taken as the starting point for comet movements. During the course of the seventeenth century, however, it was realised that this approach could not describe the motion of comets on the sky. The newly invented telescope was of assistance in this. However, the often poor optical quality did not really allow any detailed ideas about the appearance of comets to be obtained.

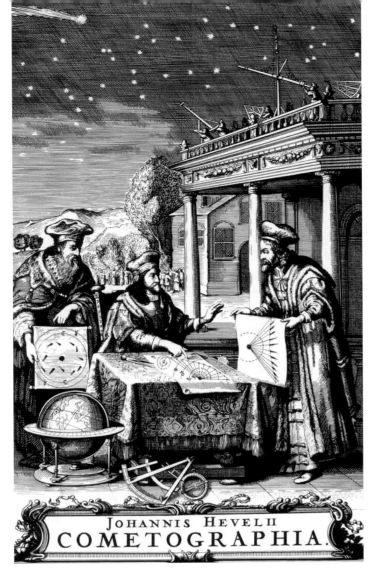

JOHANNIS HEVELII
COMETOGRAPHIA.

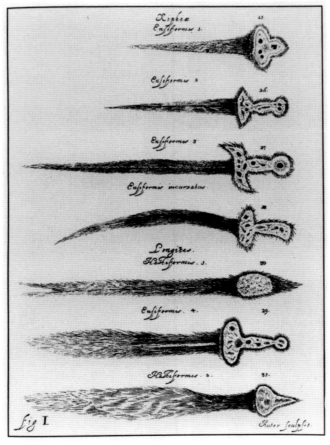

Johannes Kepler (1572–1630), who revolutionized understanding of the orbital motions of the planets, had difficulty in describing cometary orbits. He observed Halley's Comet of 1607 and, a little later, was one of the first to undertake telescopic observations of comets. He published a study *De Cometis Libellis Tres* in 1618, in which he took straight-line cometary orbits as his starting point – in contrast to the elliptical 'Keplerian' orbits of the planets. His conclusions were based on inadequate determinations of position, whereas his deciphering of the motion of the planets had been assisted by the precise position determinations by Tycho Brahe.

In the interpretation of the nature of comets, Kepler was still trapped in the old ways of thought. He believed that comets arose spontaneously out of the 'ether', and were guided by a spirit that was created and disappeared with them. Yet he rightly believed that the tail consisted of cometary particles that were dragged away from the Sun, and that sunlight was reflected from them. Kepler explained the curvature of a comet's tail as the refraction of rays from the Sun. Kepler also produced numerous astrological predictions about comets, and he was convinced that the passage of the Earth through a comet's tail would give rise to poisonous vapours.

Neither was Galileo Galilei (1564–1642), the 'father of observational astronomy', able to make any sense of comets. He thought that comets moved at right-angles to the surface of the Earth – but this could be due to the fact that most comets were first discovered at or after perihelion, so that only a portion of their orbits was observed. Galileo himself did not propose any theory of his own to explain comets, but was indefatigable in attacking his contemporaries on the subject.

The Danzig brewer and astronomer Johannes Hevelius (1611–1687), in his monumental work *Cometographia* of 1668, discussed comets thoroughly and brought the various theories that previously existed into a single overall picture. His own observations between 1652 and 1665 were included in the work, which also contained a catalogue of historical comets from the Biblical times to 1665. Like Kepler, Hevelius initially adopted a linear motion, but later established that curved paths must be involved, and that comets moved fastest when near the Sun. He took parabolas or hyperbolas as the probable form of orbit.

Hevelius supported the view that comets themselves were disc-shaped and lay at right-angles to the Sun. They arose in the atmospheres of Jupiter and Saturn by the transpiration of vapours – a reflection of the then common view that such vapours were to be found on many heavenly bodies: as spots on the Sun, for example. Hevelius chose Jupiter and Saturn on the basis of the colour of comets. The tail was formed from particles ejected from the core of the comet, and which were then swept away from the head by the Sun.

Title page (*top*) and sample page from *Cometographia*, the most influential tome on cometary astronomy in the seventeenth century. *Johannes Hevelius*

28

Title page (*top*) and sample page from *Theatrum Cometicum*, the second influential volume on cometary astronomy in the seventeenth century. *Stanislaus Lubinietzky*

At the same time as the work by Hevelius, a masterpiece by another Pole appeared. Lubinietzky was a historian who described in more than 800 pages, in his *Theatrum Cometicum* of 1666–68, the history of 415 comet appearances from antiquity to 1665. Numerous illustrations and charts meant that his book was frequently cited.

The Great Comet of 1680 brought about a considerable advance in cometary research. The German astronomer Gottfried Kirch (1639–1710) discovered it accidentally with a telescope. It was the first comet discovery made in this manner. A theologian from Plauen, Georg Samuel Dörffel (1643–1688) showed that with the comet of 1680, the path could be described by a parabola, with the Sun at the focus. His conclusion was, however, based on just a few position determinations, and it was not an accurate orbital determination. This was provided by Isaac Newton (1643–1727). He initially believed that the comet of 1680 consisted of two separate bodies before and after perihelion, which, moreover, moved on straight-line orbits. However, he rejected this idea years later in his *Principia* (*Philosophiæ Naturalis Principia Mathematica*), when he developed his method of determining orbits. This semi-graphical method allowed him to determine the shape of the orbit and the velocity of the comet from observations that were separated in time.

Newton recognized that the Sun's gravity was the fundamental cause of the motion of comets. He considered comets to be solid bodies that produced vapours because of heating by the Sun, and he explained the curvature of the tail as being caused by the comet's own motion.

Edmond Halley (1656–1742) seized on Newton's method and calculated the orbits of 24 comets seen between 1337 and 1698. In doing so, he discovered that the comets of 1531, 1607 and 1682 followed similar elliptical orbits, and therefore had to be repeated appearances of one and the same body. The differences between the periods which were not precisely equal between apparitions were correctly explained by Halley as caused by Jupiter's gravitational effects. Halley later extended the series of comet apparitions to the comets of 1456, 1380 and 1305. However the last two are incorrect, because Halley's Comet did return in 1378, but was seldom observed.

In 1705, Halley predicted a return of this comet for 1758, but later shifted the date to the turn of the year (1758–1759) because of orbital perturbations by Jupiter. In addition, he maintained that there was agreement between the comets of 1532 and 1661. He also determined a return period of 575 years for the comet of 1680, and identified it with comet apparitions in 1106, 531 and 44 BC. Whereas Halley hit the nail on the head with the comet now named after him, the second identification has, however, not proved to be correct.

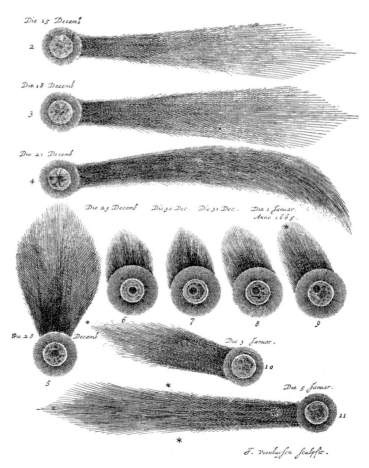

With his epoch-making identification of the comet appearances of 1531, 1607 and 1682, and his prediction for 1758, Edmond Halley revolutionized cometary research.

The cometarium

People in the eighteenth century were fascinated by the ability to calculate cometary orbits and to predict the apparitions of comets. The path of a comet around the Sun could be visualized with a cometarium, as it was known. This was a mechanical device, which demonstrated the motion of a comet around the Sun, and in particular the faster motion at perihelion. Kepler's Second Law was thus directly visible.

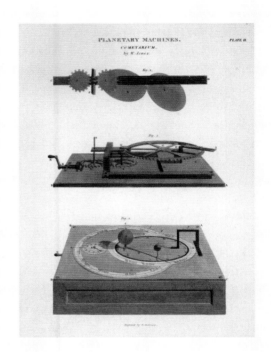

The first such device was described by the French-born, British scientist John Theophilus Desaguliers in 1730, however this depicted the orbit of the planet Mercury. The first mechanism specific for comets was built by Benjamin Martin around 1740. Between 1750 and 1850 many devices appeared, particularly by English manufacturers. Nowadays surviving cometariums are rare collectors' items.

The eighteenth century

The successful prediction of the return of Halley's Comet, thoroughly destroyed all irrational ideas associated with these celestial bodies. Comets were now seen as part of predictable celestial mechanics and as one of the most persuasive examples of Newtonian physics. In the eighteenth century, astronomers made the first deliberate searches for these celestial bodies.

The Swiss mathematician Leonhard Euler (1707–1783) refined the method of orbit determination through the use of what are now known as Euler approximations. Moreover, it was now possible to obtain the orbit of a comet, or another celestial body, from just three positions, at short intervals, with sufficient accuracy to find it again at a later date.

The Frenchman, Charles Messier (1730–1817) was the first astronomer to specifically search for comets. The French king gave him the nickname of 'The ferret of comets'. The trigger for this obsession was the search for Comet Halley that he undertook starting in 1757. The early date that was adopted (when compared with the original date), was based on the erroneous assumption that the period of Halley's Comet was decreasing. In his telescopic search, Messier came across many comet-like nebulous spots in the sky. The confusion of the Crab Nebula, nowadays designated as M1 (Messier 1), with a faint comet, gave him the impetus to compile the Messier Catalogue, which lists the brightest star clusters, nebulae and galaxies, and which is extremely popular among amateur astronomers today. Messier first found Halley's Comet in January 1759, after it had been sighted on 24 December 1758 by the Saxonian 'peasant astronomer' Johann Georg Palitzsch (1723–1788). However, Messier was not put off and went on subsequently to discover 12 comets, among them the Great Comet of 1769, which brought him international fame.

In 1797, Heinrich Wilhelm Olbers (1758–1840) published a practical method of calculating orbits, which is still used. In 1801, Carl Friedrich Gauss (1777–1855) produced yet another method.

The dishonest discoverer of comets

In the eighteenth century the discoverers of comets were honoured and rewarded. Charles Messier was allowed to present his chart of comets to the French king Louis XV in person, and following his discovery of the Great Comet of 1769, he received an increase in his salary. It was even possible for him, a commoner, to marry a noble lady. It is not surprising, then, that fraudulent imitators should appear – even those who had never discovered any comets. Jean Auguste d'Angos (1744–1833), a member of the French Académie des Sciences (Academy of Sciences) claimed to have beaten Messier in the discovery of a comet from Malta on 15 April 1784. Messier, however, was unable to confirm this. D'Angos reported additional alleged discoveries in 1793 and 1798. The latter comets he even claimed to have discovered in transit in front of the Sun.

These assertions and alleged calculated orbits were later exposed as clumsy falsifications. They were based on erroneous borrowings of data on other comets, as well as on deliberate deceit. In France, the attention given to true and false comet discoverers at the time was so great that the word 'Angosiade' became established for a fraudulent deception.

The nineteenth century

As a result of refinements in celestial mechanics, meteors were identified with the remnants of comets. From the middle of the century, a change of direction into the investigation of the structure of comets took place. The new quantitative technologies of photography and spectroscopy replaced visual observations.

Jean Louis Pons (1761–1831), with 30 confirmed comet discoveries between 1807 and 1831, was the most successful discoverer of comets in the nineteenth century. Three quarters of all discoveries in this period were to his account: between 7 August 1826 and 3 August 1827 he had found five comets by himself. Yet Pons was not a trained astronomer, but had come into contact with astronomy from being the doorkeeper at the Marseille Observatory.

Types of comet tails (*right*) and heads (*below*) based on prominent nineteenth-century examples. *Right*: Comets Donati, Coggia, the Great September Comet, Olbers and Sawerthal. *Below*: Halley, Encke, Biela and the Great September Comet.

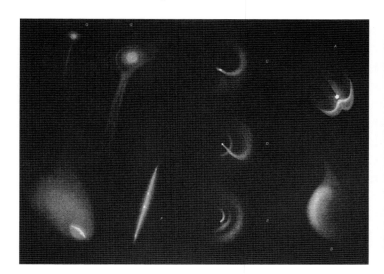

The Director of the Berlin Observatory, Johann Encke (1791–1865), calculated the orbit of the comets of 1805, discovered by Pons, and its recovery in 1818, and found an orbital period of 3.3 years, including taking into account perturbations by the planets. He predicted that it would become visible again in 1822. Today, this famous short-period comet carries his name. With further observations of 'his' comet, Encke discovered an additional

31

Biela's Comet

The nineteenth century was not only blessed with unusually many impressive bright comets, but was also the time when very mysterious events took place. One of these was related to the 'lost' Comet Biela, which caused the discovery of a comet in the twentieth century. The comet was discovered by Jacques Montaigne on 8 March 1772. In 1805, Bessel determined the period as 4.7 years. This value was later revised to 6.7 years. The next successful observation, however, was only made in 1826 by Wilhelm von Biela, a captain in the Austrian military, who (re-)discovered the comet, independently of Montaigne's observations and Bessel's orbital calculations.

The comet now known as Biela's Comet was observed at its returns in 1832 and 1845. At the end of 1845, numerous astronomers followed its break-up into what were initially two fragments. In 1852, the two portions of the comet were seen once more, but subsequently, to this very day, there have been no more observations of the comet. It remains lost – whether it broke up completely, or has been inserted into another orbit through perturbations, remains uncertain.

On 27 November 1872 the remnants of the comet gave rise to a great meteor shower (the 'Bielids') with over 1000 meteors per hour. And about 100 years later, Biela's lost comet created a furore. In 1973, the Czech astronomer Luboš Kohoutek searched for remnants of the comet with the large Hamburg Schmidt telescope, during which he initially discovered some new minor planets and subsequently found the famous and notorious Comet Kohoutek.

perturbation that created a shift in perihelion, which could not be explained by planetary perturbations. He explained this through the existence of the 'ether', which, in the nineteenth century was a widely discussed hypothetical medium between the planets.

Friedrich Wilhelm Bessel was one of the most able calculators of orbits in the nineteenth century. He also determined the perihelion advance of cometary orbits and explained this through effects from the comet itself, which was later recognized to be the correct explanation.

The Italian astronomer Giovanni Schiaparelli (1835–1910), who became known through the 'discovery' of martian 'canals', found that the orbits of meteor streams were similar to those of comets. He linked the Perseids with Comet Swift-Tuttle of 1862, and later he succeeded in the identification of the Leonids with material from Comet Tempel-Tuttle 1866.

Giovanni Batista Donati (1826–1873) and William Huggins (1824–1910) obtained the first cometary spectra of Comets Tempel 1864 and Tempel-Tuttle 1866. In doing so, they discovered three bright lines, which Huggins recognized as carbon compounds, the true nature of which was only deciphered later. (They were from molecules of CN, C_3 and C_2 that had been excited to fluorescence.) They were the first indications of the composition of comets and the first indication that comets did not just reflect sunlight, but also gave off illumination themselves.

Fyodor A. Bredikhin (1831–1904) postulated three types of cometary tail, according to the effects of the force directed away from the Sun and the mass of the particles. He assigned hydrogen gas to Type I; linked Type II to hydrogen compounds; and Type III with heavier gaseous components. He held that the different types of tail consisted solely of gases, and took the electrical charge on these as being responsible for the different forms of tail.

Barnard's automated comet search engine

Edward Emerson Barnard (1857–1923) was one of the best visual observers of his time and a pioneer in astrophotography. Like so many of the great astronomers of the nineteenth century, he started his career as an amateur astronomer. Even as an amateur he had discovered comets as early as 1881. In total, he had 16 finds, including the first photographic comet discovery in 1892, with the subsequently lost comet D/1892 T1.

On 8 March 1891, Barnard opened the *San Francisco Examiner* and read, to his astonishment, an article about a comet-discovery machine that he had apparently invented. This was supposed to scan the sky in strips and thus obtain spectra of all the objects in the field. Only when a cometary spectrum with its characteristic lines had been captured, would light fall into an apparatus that would then set off an alarm in Barnard's bedroom.

Barnard immediately tried to quash this false story, but the press had already been told that he would deny having made the invention. Numerous astronomers around the world, including famous comet discoverers such as Lewis Swift, enquired about the machine's mechanism from the increasingly indignant Barnard. Barnard fought for two years to correct these claims, until the *Examiner* finally published an apology. The identity of the originator of this canard remains uncertain, although Barnard always suspected his colleague, James Edward Keeler.

The twentieth century

As the quality of photography and spectroscopy improved, research turned to the physics and chemistry of comets. The identification of additional components of comets became possible. In the second half of the century, models explaining the processes occurring in comets and the question of their origin took centre stage.

Nicholas Bobrovnikoff (1896–1988) provided the first detailed chemical analysis of a comet in 1931, from the combined results on Halley's Comet 1910. On the basis of extensive analysis, he proved the existence of the molecules CN, CO^+, N_2^+, C_2, Na and CH.

In 1932, Ernst Öpik (1893–1985) suggested that comets originated from a cloud at the edge of the Solar System, that had formed over thousands of millions of years through perturbations of passing stars. From time to time, bodies from this cloud were dragged into smaller perihelion distances. In 1950, Jan Oort (1900–1992) took up this idea, and postulated the existence of the cloud, now named after him, of 190 thousand million cometary nuclei at distances between 10 000 and 50 000 astronomical units from the Sun.

In 1950, Fred Whipple (1906–2004) suggested the icy conglomerate model, better known by the catch-phrase the 'dirty snowball' model. In this, comets consisted of a solid icy nucleus with embedded dust. In 1981, Whipple also found a method of determining the rotation of cometary nuclei.

Kenneth Edgeworth (1880–1972) and Gerard Kuiper (1905–1973) outlined a second belt of comets between 30 and 50 astronomical units away. Kuiper derived its existence from perturbations of Neptune.

In 1951, Ludwig Biermann (1907–1986) postulated the existence of a stream of charged particles (the 'solar wind'), which is responsible for the observed accelerations and deviations from the anti-solar direction of the gas tails of comets. Comet tails, in particular the gas tails, are thus indicators of the direction and velocity of the solar wind.

Towards the end of the century, the first comet discovery from Earth orbit was made by the Solwind satellite in 1979. The Spacewatch programme provided the first time a comet was discovered by an automated search system (125P/Spacewatch). The first direct images of a cometary nucleus were obtained by the Soviet Vega 1 and 2 space probes and, above all, by the European Giotto probe (at a distance of just 600 km) during the passage of Comet Halley in 1986.

The twenty-first century

Since the turn of the century, research into comets from space has been in the forefront. Space probes have impacted on comets, have observed them from nearby, and will soon land on them.

In 2005, the Deep Impact probe struck Comet 9P/Tempel with an artificial impactor. The impact was directly observed from the mother probe. The first landing on a comet (67P/Churyumov-Gerasimenko) is expected to take place with the Philae lander from the Rosetta space probe.

Comet discoverers

The search for comets has repeatedly fascinated patient people, not least because the celestial objects bear the name of their discoverer.

In China, a systematic search for comets with the naked eye was carried out for 2000 years. The discoverers, court astronomers to the Chinese emperor, however, remain largely unknown. In Europe,

Nowadays comet research primarily takes place in space. While Giotto obtained the first direct image of a comet's nucleus in 1986 (*left*), Rosetta will attempt to land on a comet in 2014 (*below*).

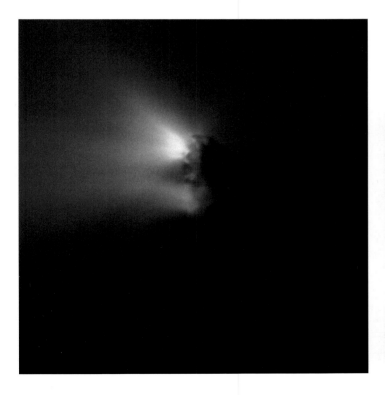

Halley's Comet

Halley's Comet – more correctly known as 1P/Halley – occurs as one of the Great Comets of modern times on no fewer than seven occasions. Not every return has actually produced an impressive sight in the sky, but every return since 1500 has been a milestone in cometary research.

- In 1531, Peter Apian recognized that the tail always pointed away from the Sun
- In 1607, Johannes Kepler succeeded in obtaining an orbital calculation (albeit an erroneous one)
- In 1682, Edmond Halley observed the comet named after him, which later led to his famous prediction of its return
- In 1759, the first systematic search for a comet was undertaken, and Johann Georg Palitzsch succeeded before Charles Messier
- In 1835, Wilhelm Bessel calculated the size of the mass-loss at every return
- In 1910, a worldwide photographic monitoring programme was undertaken for the first time, in which more than 60 observatories on every continent took place
- In 1986, there was the first visit by space probes to a comet, in which European, Japanese and Soviet missions all took part.

A total of 30 apparitions of Halley's Comet have been documented since 240 BC. It is, however, believed that the comet has existed in its current orbit for at least 16 000 years, and possibly for as many as 170 000 years. Because of orbital perturbations, however, it is not possible to carry calculations backwards beyond 1404 BC. The orbital period averages 76 years, but varies between 74.9 and 79.2 years because of perturbations by the giant planets, above all, by Jupiter.

Despite its great age and its many orbits of the Sun, the comet is still very gas-rich. Despite a mass-loss of 3×10^{11} kg per orbit, no decrease in the gas-production rate can be determined. The overall mass of the nucleus, which is 15 km × 8 km × 8 km in size, is given as 2.2×10^{14} kg. According to estimates, the comet has exhausted about half of its potential. This enormous reservoir has resulted it in being the only periodic comet that is visible to the naked eye at every return. Remnants of the comet's tail form the Eta Aquarid and Orionid meteor showers, which occur when the Earth crosses the comet's orbit, which is inclined at 162° to that of the Earth.

Its orbit takes Halley's Comet from a distance of 523 million kilometres (35.1 AU) beyond Neptune, to just 90 million kilometres from the Sun. While its next aphelion passage occurs on 9 December 2025, we will have to wait almost 50 years for its next return – the comet will only be at perihelion on 28 June 2061.

The attractiveness of its appearance from Earth is primarily determined by its distance from us. This may vary between 0.5 and 0.03 astronomical units (75 to 4 million kilometres). The closest pass in historical times was the return of 837. The next close approach will occur in 2134, at a distance of 0.09 AU. The reflectivity (albedo) of the comet is, at 4 per cent, comparable with that of coal – the comet is almost completely black.

Summary of cometary missions (after Leitner/Pilz)

Name	Target comet	Country/Organisation	Launch	Date of encounter	Distance of encounter
ISEE-3/ICE	21P/Giacobini-Zinner 1P/Halley	USA/NASA	12 Aug. 1978	11 Sep. 1985 28 Mar. 1986	7800 km 31 000 000 km
Vega 1 Vega 2	1P/Halley	USSR, Austria, German Federal Republic, France	15 Dec. 1984 21 Dec. 1984	6 Mar. 1986 9 Mar. 1986	8890 km 8030 km
Sakigake	1P/Halley	Japan	8 Jan. 1985	11 Mar. 1986	7 000 000 km
Giotto	1P/Halley 26P/Grigg-Skjellerup	ESA	2 Jul. 1985	14 Mar. 1986 10 Jul. 1992	596 km 200 km
Susei	1P/Halley	Japan	18 Aug. 1985	8 Mar. 1986	151 000 km
Deep Space 1	19P/Borelly	USA/NASA	24 Oct. 1986	22 Sep. 2001	2200 km
Stardust	81P/Wild 9P/Tempel	USA/NASA	7 Feb. 1999	2 Jan. 2004 14 Feb. 2011	240 km 181 km
Deep Impact EPOXI	9P/Tempel 103P/Hartley	USA/NASA	12 Jan. 2005	2 Jul. 2005 11 Oct. 2010	500 km 700 km
Rosetta	67P/Churyumov-Gerasimenko	ESA	2 Mar. 2004	Nov 2014	Landing

The Kreutz Sungrazers

In 1888, the German astronomer Heinrich Kreutz noted that the orbits of comets that passed particularly close to the Sun, were very similar. He correctly understood this as a sign of a single, common parent object for these 'sungrazers'.

The objects now known as Kreutz comets form what is called a 'family' of comets: the members move on similar orbits around the Sun. Perihelion lies at a distance of just 0.005 AU to 0.009 AU from the centre of the Sun, which corresponds to distances of between 0 and 700 000 kilometres from the surface of the Sun – many Kreutz comets impact directly onto the Sun, and most break up during their extremely close passages to the Sun. Their orbital periods amount to 500 to 1000 years; the semi-major axis of the orbit is 150 AU; the orbital inclination is about 140°; and the longitude of the ascending node is about 0°.

Over 2000 Kreutz comets are known. Most of these bodies are very small (with diameters of up to 130 m) and do not survive perihelion passage. However, shortly before that, they may become very bright and may even appear close to the Sun in the daytime sky.

The majority of sungrazers are discovered by amateurs on images from the SOHO solar probe, when they are close to the Sun. Kreutz comets thus make up more than 80 percent of all cometary discoveries. A few amateurs have found several hundred comets in this way.

Because of their extremely close approaches to the Sun, a few Kreutz Sungrazers have become exceptionally impressive in appearance. Among these are the Great March Comet of 1843, the Great September Comet of 1882, and Ikeya-Seki in 1965. In 2011, the Kreutz comet C/2011 W3 (Lovejoy) appeared, which survived its passage past the Sun and developed a long tail after perihelion, but which was visible only from the southern hemisphere for a short period.

The comet C/2011 W3 Lovejoy was the last bright Kreutz Sungrazer. It survived solar passage, which took it just 140 000 km from the surface of the Sun.

According to Brian Marsden, the origin of the Kreutz family of comets is possibly to be found in a comet that appeared in 372 BC, as well as (for the comets of 1882 and 1965) once again in 1106. Subsequently, the fragments from the disintegration of these two comets formed the bodies that are now part of the Kreutz group. Alternatively, according to more recent studies by Sekanina and Chodas, the apparitions of comets in 214 BC and AD 467 are responsible.

As well as the Kreutz group of comets there are other families of sungrazers: the Meyer, Marsden and Kracht groups. They, however, do not produce notable tails when they come close to the Sun and are short-period objects.

---

until the seventeenth century, comets were only discovered if they were generally visible. As a result, here too, practically no discoverers can be determined. That then changed in the eighteenth century, in particular through the search for Halley's Comet at its return in 1758–59.

Kirch made the first telescopic discovery in 1680, and the first comet to be discovered photographically was by Barnard in 1892. If comets discovered to date are taken as a whole, the first four places in a list of the greatest comet discoverers go to modern automated searches that scan the skies for comets. The leading actual person is the Scottish-Australian astronomer, Robert McNaught, with 82 discoveries – although these were all made photographically.

Nowadays, very few comets are discovered visually, because the

The most successful comet search programmes (as at 5 Sep. 2014)	
SOHO	1979
LINEAR	215
Catalina Sky Survey	123
Siding Spring Sky Survey	101
Mount Lemmon Survey	63
Spacewatch	61
PanSTARRs	61
NEAT	54
LONEOS	42
STEREO	42

automated searches and astronomers working with photography – the latter almost exclusively amateurs – are able to detect considerably fainter objects. In general, these discoveries are accidental, as with the German astronomer Sebastian Hönig, who found a comet visually in 2002 with a 10-inch telescope. Worldwide, however, only twelve comets have been found visually since 2000. At present, the leader is the American, Don Machholz.

If the visual and photographic comet discoveries by amateurs are taken together, then in recent years between two and eight comets have been discovered every year. This figure, however, conceals the fact that a large number of comets found by the SOHO solar observatory are actually the result of hard-working amateurs, who evaluate the images returned from space in real time, and have thus been able to detect more than 1000 new comets.

The most successful comet discoverers with at least ten discoveries (data from Maik Meyer)

Discoverer	Total	As sole discoverer	Visual discoveries	Photographic/CCD discoveries	First discovery	Last discovery
Robert H. McNaught	82	70	0	82	C/1978 G2	C/2013 E1
Carolyn & Eugene Shoemaker	32	13	0	32	C/1983 R1	P/1994 J3
Rik E. Hill	27	23	0	27	P/2004 V5	C/2014 F1
Jean-Louis Pons	26	22	26	0	C/1801 N1	C/1827 P1
Andrea Boattini	25	22	0	25	C/2007 W1	C/2013 J5
Alex R. Gibbs	23	23	0	23	C/2006 U7	P/2012 K3
David H. Levy	22	6	9	13	C/1984 V1	P/2006 T1
Eric J. Christensen	22	20	0	22	P/2003 K2	C/2014 M2
William R. Brooks	21	18	21	0	C/1883 D1	C/1911 O1
William A. Bradfield	18	18	18	0	C/1972 E1	C/2004 F4
Gordon J. Garradd	17	16	0	17	C/2006 L1	C/2010 H1
Edward E. Barnard	16	13	15	1	C/1881 S1	P/1892 T1
Brian A. Skiff	16	11	0	16	D/1977 C1	C/2007 H2
Jean Mueller	15	14	0	15	P/1987 U2	P/1998 U2
Malcolm Hartley	13	4	0	13	P/1982 C1	C/1999 T1
Antonin Mrkos	13	8	11	2	C/1947 Y1	P/1991 F1
Lewis Swift	13	8	13	0	P/1862 O1	C/1899 E1
Michel Giacobini	12	10	12	0	P/1896 R2	P/1907 L1
Eleanor F. Helin	12	2	0	12	C/1977 H1	P/1993 K2
Minoru Honda	12	5	12	0	C/1940 S1	C/1968 Q2
Charles Messier	12	11	12	0	C/1760 B1	C/1798 G1
Kenneth S. Russell	12	5	0	12	P/1979 M2	C/1996 P2
Wilhelm L. Tempel	12	10	12	0	C/1859 G1	C/1877 T1
Donald E. Machholz	11	9	11	0	C/1978 R3	C/2010 F4
Leslie C. Peltier	10	3	10	0	C/1925 V1	C/1954 M2
James V. Scotti	10	10	0	10	P/2000 Y3	P/2013 A3
August Winnecke	10	6	10	0	C/1854 Y1	C/1877 G1

Comet science today

Comets are small Solar-System bodies. Currently, about 5000 individual comets are known. However, in the outer regions of the Solar System, far from the light of the Sun, it is believed that there may be as many as 10^{12} comets. They are, however, too far from the Earth to be observed. Nowadays, comets are generally first discovered if they enter the inner Solar System within the orbit of Jupiter.

Orbits

Comets, like all other bodies, follow Keplerian orbits around the general centre of gravity of the Solar System, inside the Sun. The orbits are, in general, extreme ellipses. Near the Sun, comets move most rapidly, in accordance with Kepler's Second Law. Hyperbolic orbits are also possible, and these objects will leave the Solar System after pas-

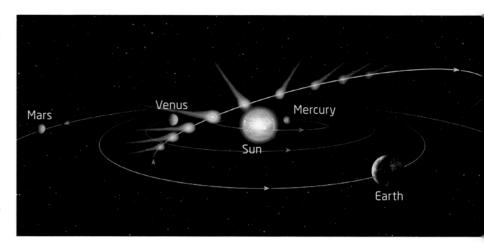

As a comet approaches the Sun, its tail develops. This points in the opposite direction to the Sun.

Comet designations

In the modern, internationally recognized and unified nomenclature, every comet, irrespective of its common name, has an alphanumeric code, which identifies it unambiguously. This alphanumeric code is assigned according to a fixed scheme. It consists of a letter prefix, a year number as well as a letter-number combination for long-period comets (with orbital periods greater than 200 years), as well as the prefix and a name for short-period comets.

The prefix indicates the status of the comet. 'P' stands for short-period comets; 'C' for long-period comets; 'D' indicates a lost or destroyed object; the prefix 'X' means that the orbit has not been determined; while 'A' designates a minor planet, that is, one that is not out-gassing. The latter have their own nomenclature.

After the second observed return of a short-period comet it is given the name of the discoverer, or the first person to calculate the orbit, as well as a number, indicating its position in the sequence of discoveries. This is in the form of 'number, P/discoverer'. As a result, Halley's Comet is officially known as 1P/Halley. Long-period comets have a detailed code. The year-number indicates the year of discovery. The following letter indicates the half-month of the year in which the comet was discovered. 'A' therefore covers the period from 1st to 15th January. As 'I' is omitted, 'Y' indicates the last half-month between 16th

and 31st December. The subsequent number then indicates the position of the comet in the series of discoveries in the corresponding half-month. Finally, after the code, the name of the discoverer is given in parentheses. Before this modern nomenclature was introduced in 1995, another system was used. After discovery, the comet was given a provisional designation, in which the year of discovery was followed by a lower-case letter, indicating the sequence of the discovery in the yearly series. If more than 26 comets were found in any one year, the letter sequence was extended with the numbers 1, 2, etc. After the orbit had been calculated, the comet was given a final designation, which consisted of the year of perihelion, and a Roman numeral indicating the sequence of perihelion passage.

Comet Bennett was given the following designations in these systems:
- modern designation: C/1969 Y1 (Bennett)
  = a long-period comet, discovered in the second half of December, 1969
- old provisional designation: 1969i
  = the ninth comet discovered in 1969
- old final designation: 1970 II
  = the second comet to reach perihelion passage in 1970.

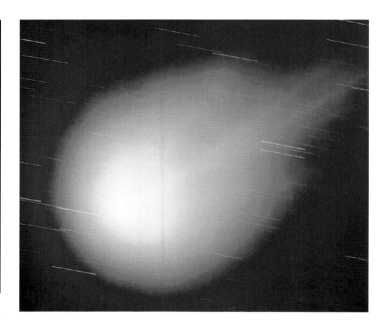

The orbit of Halley's Comet is a typical example of a cometary orbit.

sing around the Sun. Only about 4 per cent of all comets exhibit such an orbit, which probably arises through gravitational perturbations of the small cometary body by the more massive planets. Comets from other stellar systems have not been detected.

Rather less than 500 comets return within 200 years. They are known as short-period comets. Most of them actually originated from the outer Solar System, but were 'captured' by the gravitational influence of the giant planets, and injected into short-period orbits. Most have undergone many returns; the most famous example is Comet Halley, which has an orbital period of about 75 years.

Depending on which planet captured the comet, different so-called families of comets have been created. The comets in such families have similar distances when they are farthest from the Sun (at aphelion). The Halley- or Neptune-family contains about 60 comets, the Jupiter-family as many as 400. Only a few comets have orbits that keep them always within the inner Solar System. The comets known as 'Centaurs', in contrast, always remain between the orbits of Jupiter and Neptune.

Long-period comets have orbital periods of at least 200 years, although most are significantly greater. It is therefore highly probable that they are arriving 'fresh' from the outer Solar System, and have undergone the journey towards the Sun only a few times. If they approach the latter very closely, they are known as 'sungrazers'. Because of the Sun's tidal effects it may happen that when a comet passes close to the Sun, it breaks up or is completely destroyed. In addition, many comets fall directly into the Sun. Over a long time-scale, most comets end in this manner, others leave the Solar System altogether, break up, or hit one of the planets.

Long-period comets probably come from the Oort Cloud, an unobservable reservoir of small bodies at the edge of the Solar System, some 10 000 to 50 000 times farther from the Sun than the Earth. They are steered into an orbit within the inner Solar System by gravitational perturbations. In this, neighbouring stars outside the Solar System that pass close by may play a part. It is thought that there are at least one thousand million comets in the Oort Cloud.

In addition, there is also the Kuiper Belt, which, at a distance of between 30 and 50 astronomical units, is much closer to the Earth. It is believed that about 100 million comets are to be found there.

The nucleus

The nuclei of comets are, in comparison to the other bodies in the Solar System, extremely small. They range from fifty metres to a few dozen kilometres across. Sizes between one and five kilometres apparently occur frequently. But there are larger lumps: Comet Halley amounted to about 15 km × 8 km × 8 km. It is not spherical, but shaped rather like a potato – which is a property shared by many other comets. The mass of a typical comet lies between one million and a few million tonnes – which is not great by cosmic standards.

The actual nucleus of a comet is just a few kilometres across. It consists of frozen loose material. It turns to vapour through the effect of solar radiation and the comet outgasses. This happens only on a few parts of the surface.

The innermost shell of dust and gas around the nucleus is known as the coma. It may reach a few tens of thousands of kilometres in diameter.

There are numerous models that describe the composition and structure of a comet's nucleus. Among these is the ice-conglomerate model by Fred Whipple, which has come to be known as the 'dirty snowball'. Other models are the conglomerate, the 'rubble pile', and the ice-cemented models. Nowadays, it is generally believed that there is a greater amount of dust than in Whipple's original model. The dust is held in place by the ice. So it is probably more correct to speak of an 'icy dustball'.

Comets are darker than asphalt, and reflect only about 3 per cent of the incident sunlight. At the same time they have a very low density and are lighter than water. Dust particles primarily contribute to the mass, and their sizes lie between a micron (10^{-6} m) and a centimetre. Silicates dominate, but there are also carbon, hydrogen, oxygen and nitrogen compounds. Particularly notable is the presence of long-chain molecules. These substances, known in chemistry as organic molecules, such as glycine (an amino acid: $C_2H_5NO_2$) may be detected in comets. Such molecules are the building blocks of hydrogen/carbon-based life as we know it. As 'fresh' comets from the Oort Cloud are more-or-less pristine, because they are essentially unaffected by any chemical or geological processes, but are, at the same time, extremely old bodies, it is believed that the primordial material from which the Solar System formed contained these amino acids. It is probable that comets first brought these compounds to Earth, and thus created the conditions for the origin of life.

The ice component that contains the dust consists of about 80 per cent water ice, 15 per cent carbon monoxide, and 4 per cent carbon dioxide. In addition, traces of methane, ammonia, and nitrogen are present as ices.

The coma

At great distances from the Sun, cometary material is frozen solid. The comet is inactive, and it consists of just its nucleus and, as such, differs little from a minor planet. A comet's approach towards the Sun brings about the difference. Individual ice components begin to turn into vapour (to sublime) with the increase in solar radiation. A shell of gas forms around the nucleus, known as the coma (from the Latin for 'hair'). At a distance of 11 astronomical units (1.6 thousand million kilometres) cyanide begins to sublime. At 6.5 astronomical units, carbon monoxide is released and ammonia at 6 astronomical units. Hydroxyl (OH) radicals appear at 3 astronomical units as well as the lion's share of water ice. Together, the fractions that have sublimed form a gaseous shell 10 000 to 100 000 kilometres in diameter, and in some cases, significantly larger. A hydrogen shell may reach a diameter of up to 100 million kilometres.

The fraction of volatile material varies from comet to comet and is also dependent on the orbital period of the comet and the structure of its crust. The latter's structure also affects the release of the dust. The dust particles are released and transported through the sublimation of the ice. However, this probably occurs only in strictly

The tail of a comet may extend for many million kilometres. It consists of gas molecules (here appearing blue) and dust particles that leave the region around the nucleus. *Gerald Rhemann*

Kepler's Second Law

The line between a planet and the Sun (the 'radius vector') sweeps out equal areas in equal times. This fundamental statement applies to all other bodies in the Solar System.

This can be understood particularly well when a comet comes close to the Sun. The closer the comet to the Sun, the faster its motion. The velocity peaks when closest to the Sun, at perihelion. As the comet retreats from the Sun, its velocity decreases. It is thus proportional to its distance from the Sun.

The 'ingredients' for a great comet

Every year about a dozen comets come close to the Sun and the Earth. Most of these objects, however, are visible only to amateur and professional astronomers through telescopes. At about yearly intervals a comet may be observed with binoculars. With the naked eye, a comet is visible only about once every two to three years. Bright comets, that can be recognized immediately in the night sky, are, in contrast, visible about once a decade, at the most. A few fortunate circumstances must combine for a great comet to appear:

- The orbit must give a close approach to Earth as well as a close approach to the Sun. In most cases, only one of these factors is present.
- Favourable conditions occur when the comet is near perihelion and is visible in a dark sky – although because it is then near the Sun, the comet is generally only seen in twilight.
- A long duration of visibility significantly increases attention on the comet – frequently, however, the phase of greatest brightness is only a few weeks or even just days long. It is particularly pleasant when the comet is visible in the evening – many bright comets have been seen by just a few early risers.
- If the comet itself has a large nucleus, and is accompanied by a lot of fresh material, then a greater magnitude is to be expected. An elevated production of dust – which differs from comet to comet – allows a dust tail to arise, and this contributes most of the brightness. This may also rise sharply through an outburst on the comet, that causes a sudden increase in the release of material.
- The discovery of comets must also take place a sufficiently long time before the brightest phase. Before the existence of the Internet, news of the appearance of a comet in the sky was difficult to spread in an adequate time. The date of discovery also determines the general level of attention given, but which is also affected by other factors – whether, for instance, a comet falls within the summer holidays or is expected around Christmas. By no means last, the weather also plays a part.

However, humanity itself has largely destroyed one of the most important factors governing the occurrence of an impressive apparition of a comet: a dark night sky. Many of the greatest comets go unseen by most people, because they are drowned out by light pollution from towns and cities. On this point, Fred Whipple remarked "I am not sure that there was ever a comet bright enough to appear spectacular from downtown New York."

limited areas that are turned towards the Sun, and which cover just a fraction of the surface. Fountains of ice and dust, known as jets, are created. The rotation of the comet nucleus causes them to act like lawn-sprinklers and to spread dust into the coma.

The sublimation of icy fractions cools comets. At the same time, the gas in the outer coma is dragged away by the solar wind, the stream of charged particles from the Sun, as well as by its radiation pressure.

The tail

The characteristic feature of comets is a tail. It first forms when the comet arrives within the inner Solar System, inside the orbit of Mars.

Two types of tail are differentiated from their spectra: The first type consists of molecules that are drawn out of the coma by the solar wind. This is also known as the gas or plasma tail. It is generally long and narrow, and always points directly away from the Sun. Its length may amount to as much as 100 million kilometres, which corresponds to two-thirds of the distance of the Earth from the Sun.

Through its interaction with the solar wind, the gas tail illustrates the Sun's magnetic field. Alterations in the density of the field or changing polarity can thus lead to tail disconnection events, when the flow of molecules being dragged away from the comet is broken, only to reform shortly afterwards.

The second type of tail consists of microscopic particles of dust. They are expelled from the inner coma by radiation pressure from the Sun. In this, the mass of the particles plays a part. It amounts to a selection or differentiation process, in which the mass-poor particles follow different paths to the mass-rich ones. In contrast to the gas tail, they thus trail behind the comet in a different way. This is one of the reasons for the creation of the typical curved shape of dust tails, which do not precisely form in the opposite direction to the Sun.

Dust tails form only with copious dust production. This usually occurs only with great comets or after perihelion. If a lot of dust is released all at once, then rays occur in the tail, which are known as 'synchrones'. If such events recur, for example as a result of rotation of the cometary nucleus, then many such rays may appear, and the-

The visual magnitude of comets

The overall magnitude of a comet gives information on how well it will be seen. In doing so, however, one must differentiate between conditions with a dark night sky (without any light pollution) and with the various phases of twilight:

- < −8: The comet is clearly visible in daylight. Such magnitudes are also only possible close to the Sun.
- −8 to −4: With knowledge of its position, the comet may also be seen in the daytime sky, if the Sun is obscured. The comet is conspicuous in twilight and highly striking in the night sky.
- −4 to 0: The comet first appears to the naked eye in deep twilight. In the night sky its appearance is striking.
- 0 to 2: The comet is difficult to see in twilight. In the night sky it is clearly visible to the naked eye.
- 2 to 4: The comet is invisible in twilight. In the night sky it may be detected with the naked eye, when its position is known.
- 4 to 6: Only experienced observers at dark sites will succeed in seeing the comet without any optical aid. However, the comet is clearly seen in binoculars.
- 6 to 8: The comet cannot be seen with the naked eye. It appears faint in binoculars.
- 8 to 10: Comets of this magnitude may only be seen through small amateur telescopes.
- 10 to 12: A large amateur telescope is required for these comets.
- > 12: These comets are generally only detectable through photography.

The apparent size of comets

Faint comets generally appear to the naked eye as star-like or like a fuzzy star. The length of the tail must amount to at least 1° for it to be clearly seen by the naked eye. Shorter tails and the coma are normally visible only with binoculars or a telescope.

The star-like or very small nucleus visible with such instruments is not the true cometary nucleus, but the innermost region of the coma. It is known as the false nucleus. In historical reports it is, however, generally described as the nucleus. The first view of the actual nucleus of a comet was obtained by the Giotto space probe, which passed Halley's Comet in 1986.

se are known as 'striae', (from the Latin for 'furrows'). Dust tails may also reach lengths of up to 100 million kilometres.

What might seem to be a third type of tail, known as an anti-tail, may also occur. This, unlike the two classic types, points towards the Sun rather than away from it. This is, however, just a perspective effect, which occurs when the Earth lies in the comet's orbital plane, and the curvature of the tail causes it to appear in the wrong direction.

The larger particles from a comet's tail become distributed along the comet's orbit as dust trails. If the Earth subsequently crosses such a trail, numerous meteors are seen to streak through the night sky. Most of the meteor showers that recur every year may be ascribed to specific comets. There are suggestions that this accretion of cometary material could have provided the basic components for life on Earth in the form of carbon-based macro-molecules.

Great comets in antiquity

Since primeval times, the appearance of bright comets has impressed humanity. But more detailed conclusions about individual comets or even their paths can only be obtained from written records, which are available from about 500 BC. The most complete reports of comets come from China, whereas reports from Europe are incomplete, and only single out certain individual comets. Chinese astronomers also gave details of the oldest cometary apparition that can be identified nowadays, from the year 613 BC. Readily understood indicators are provided by Halley's Comet, which was observed in the years 240, 164, 87 and 12 BC. After the birth of Christ, it appeared in AD 66, 141, 218, 295, 374 and 451.

One of the oldest comets documented from Europe is the Great Comet observed by Aristotle in the winter of 373–372 BC. The Greek thinker saw it in his childhood. He later wrote in his work *Meteorologica*: "its light stretched as a great ribbon across one third of the sky". Aristotle linked the comet with an earthquake and with a 'tidal wave' that the latter produced – an early example of the interpretation of comets as harbingers of disaster that persisted into modern times. According to Diodorus, witnesses claimed that the comet had cast shadows like the Moon. A few chroniclers report that the comet eventually 'broke into two planets'. According to modern analysis, it was probably a sungrazer, which broke up after a close approach to the Sun. If this is so, the Comet of 372 BC may be the parent comet of the Great Comets of 1680 and 1843.

Dating from the year 240 BC, the first full sightings of Comet Halley come from China and Mesopotamia. At its return in 164 BC the apparition of the comet was documented on Babylonian cuneiform tablets. The comet of 135 BC is famous for appearing at the birth of King Mithridates VI of Pontus. It was visible for 70 days; Seneca described it, according to eye-witnesses, as "stretching out like the Milky Way". Mithridates later had coins struck with a comet as motif – a tradition that persisted until modern times. Because of its appearance, the comet was likened to a horse's mane, and was also called the 'horse comet'.

The comet that appeared after the death of Julius Caesar in 44 BC has also gone down in history. The comet, known as the 'Julium Sidus' ('Julian Star'), appeared during the eleven-day games of Venus Genetrix. The goddess symbolized the mother of the Roman people, and was the patroness of the Julian family. Caesar himself had introduced the games in 46 BC. Many contemporaries saw the comet as a torch, marking the murder of Caesar, and thought that it represented his soul, rising to heaven. Shakespeare transferred the appearance of the comet to a time before Caesar's death and thus used its appearance as an announcement of his death. The comet was also seen in China, where the annals report the length of the tail as up to 15°. However, the Chinese observations show discrepancies in timing when compared with those from Rome.

The appearance of Halley's Comet in 12 BC served to announce the death of the Roman commander, Marcus Agrippa. In China it was followed for more than 56 days. It was the source of 'a great deal of worry' for the Chinese emperor, because he took the comet as an omen of an attack and warned his officials to be vigilant.

The brightest comets of the first century AD all became famous as signs of death and troubles. According to Chinese chroniclers, in June–July 54, a bright comet was seen in the sky for about a month. It was taken to be a sign of the death of the Roman emperor Claudius, who was poisoned by fungi by Nero's mother Agrippina in October of the same year.

Even more famous was Nero's Comet of AD 65. According to Seneca it was visible for six months from Rome, and the Chinese astronomers recorded it for 135 days. During this time the then emperor, Nero, committed one of his acts of tyranny; fearful of the end of his rule, he ordered the deaths of some of the Roman ruling classes, and had others send into exile. Tacitus, writing in his *Annals*, said: "A comet appeared, upon which Nero took care to shed the blood of some outstanding person as an expiatory sacrifice." The assassination of his wife Octavia in AD 62 and Seneca's suicide on Nero's orders in AD 65, were also linked to comets – although the appearance of a second comet was possibly involved there. Seneca described comets as 'malevolent torches' in his play about Octavia's downfall. He was, however, opposed to the astrological interpretations of comets. According to legend, the comet was predicted by Aristotle and Theophrastus.

Comet Halley's appearance followed just one year later. It was linked to the destruction of Jerusalem by the Romans. The Romano-Jewish historian Flavius Josephus described the comet as 'a star like a sword' that hung, menacingly, over Jerusalem. It is possible, however, that another apparition of a comet in AD 70–71 is intended. In AD 79, another comet appeared on the scene, which was interpreted as a negative sign for the Emperor Vespasian. According to tradition, however, the latter shrewdly held that the hairy nature of the star would be more appropriately applied to the Persian rulers than to his bald head. Nevertheless, Vespasian died in June of that same year.

Comet Halley's return in AD 218 was seen by contemporaries as a sign of the uprising against Emperor Opellius Macrinus. The comet 'caused great apprehension' and was visible for 40 days during May and June. The second-closest approach to Earth by Comet Halley in historical times took place in AD 374, when the separation was only 13 million kilometres. There are, however, no European observations of this return, and only a few reports are known from

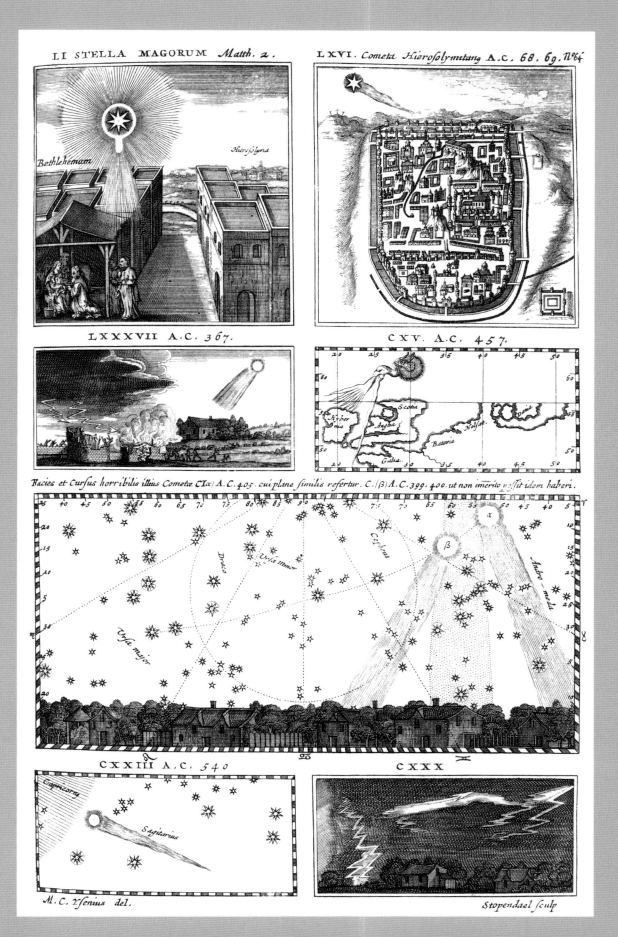

Great Comets since the birth of Christ. In the first image, the Star of Bethlehem is depicted. From the *Theatrum Cometicum* of *Stanislaus Lubinietzky*.

The Star of Bethlehem

Generations of theologians and astronomers have sought a meaning in the famous passages in the second chapter of the Gospel according to Matthew, which deal with the three holy kings and the birth of Jesus. The text reads: "Now when Jesus was born in Bethlehem of Judaea in the days of Herod the king, behold, there came wise men from the east to Jerusalem, saying, Where is he that is born King of the Jews? For we have seen his star in the east, and are come to worship him." (Matthew 2:1–2) The three wise men travelled to Bethlehem, following Herod's advice. "When they had heard the king, they departed; and, lo, the star, which they saw in the east, went before them, till it came and stood over where the young child was." (Matthew 2:9)

It was 200 years after these events that a comet was suggested as the heavenly body in question. According to the modern view, however, the Star of Bethlehem was probably not a comet – it also does not appear credible because comets were then widely seen as signs of impending disaster. Johannes Kepler in the seventeenth century suggested that a triple conjunction of the planets Jupiter and Saturn in the year 7 BC should probably be interpreted as the Star of Bethlehem. But it is also possible that we might be dealing with another close conjunction of planets, where the planets appear as a single 'star' to the naked eye, or else with the appearance of a nova or super-

nova. In all seriousness, to the modern point of view, it is impossible to know, with any degree of certainty, whether we are dealing with an actual astronomical event, or with a legend created subsequently by the evangelists.

The closest appearance in time of a bright comet was the return of Halley's Comet in 12 BC. However, its visibility is too early, according to current views of the date of Christ's birth, which is taken to be between 7 and 4 BC. In addition, according to Chinese sources the comet was observed during the late summer and autumn, from 26 August until 20 October, which does not agree with the traditional date of Christmas. It is unlikely that any other comet could have been the Star of Bethlehem, because there are no such indications from the otherwise reliable Chinese astronomers.

In the Middle Ages and in early modern times, although the scene of the three kings and the Christ child was painted frequently, only in a few cases is a comet depicted. The negative astrological connotations made such an association unlikely. Only after the fear of comets had been overcome were comets introduced as a motif in scenes of the Adoration of the Magi. The modern-day popular depiction of the Star of Bethlehem as a comet should therefore been seen as purely symbolic.

China. In AD 390, a great comet appeared in the sky, to which an extremely close encounter with the Earth of about 14 million kilometres was also ascribed. Chinese sources report tail lengths of up to 100°. The Greek historian Philostorgios described it as 'hardly any fainter then Venus' and compared its shape to that of a sword. It appeared like a pillar to the Roman historian Ammianus Marcellinus and was visible for 30 days.

Another sword-shaped comet created a sensation in AD 400. It appeared 'larger than anyone had ever seen before', and its closest approach to Earth possibly amounted to just 10 million kilometres.

In AD 442 a bright comet was followed for more than 100 days. Its perihelion lay outside the Earth's orbit, and so it could be seen in the night sky in opposition to the Sun. Halley's Comet at its return in AD 451 was clearly visible in the morning sky, when, partly during the night, the great battle of the Catalaunian Plains took place, in which Attila was stopped by the Romans. In AD 467, according to the Byzantine chronicler Theophanes, a comet appeared 'in the form of a trumpet' for 40 days. This rounds off the list of great comets in antiquity.

Great comets in the Middle Ages

During the Middle Ages, the appearances of Comet Halley marked the passage of time: It was in the sky in the years 530, 607, 684, 760, 837, 912, 989, 1066, 1145, 1222, 1301, 1378 and 1456. Many of its apparitions were very bright and were well-observed in Europe and Asia.

The return in the late summer of 530 was preceded, on 9 April of the same year, by a great meteor shower, when the Earth passed through the orbit. This produced a major outburst of the Eta Aquarid meteor shower, which may be observed every year in April and May. In the *Nuremberg Chronicles*, the appearance of Comet Halley in 684 was held to be responsible for 'great rains and lightning strokes' with injuries and deaths in Italy and Rome. In China the comet was seen for 33 days. Korean and Japanese observations have also been preserved with those from China.

At Easter in 837, Comet Halley made its closest approach to Earth in recorded times. It was less than 5 million kilometres from the Earth. That corresponds to about 13 times the distance to the Moon. At this return the tail was well over 120° long, and the comet's overall magnitude was −4 mag. It was observed from March to May, during which time the tail pointed in every direction in succession on the sky. According to legend, Louis the Pious took the comet as a sign for a return to God, although his chronicler Aginard said to him "Do not fear signs in the heavens, they only frighten fools." Just one year later, in November and December, another 'dreadful' comet appeared, which 'stretched across the whole sky'.

In 905, a very bright comet appeared in the daytime sky from 18 May. Chroniclers reported 'a terrible, large, and impressive comet, which terrified everyone that saw it'. In May and June, after perihelion, it was in the evening sky, and its tail reached 100° in length. Comet Halley's return in 1066 was taken as an omen for the conquest of England by William the Conqueror, who defeated his adversary Harold at the Battle of Hastings on 14 October 1066. The comet is shown on the famous Bayeux Tapestry, which depicts events before and after the battle (see page 17). The comet was visible in the morning sky from 3 to 22 April, and in the evening sky from 24 April to 6 June. Perihelion had already occurred on 21 March. Many sources report an outburst of the comet on 24–25 April – an event that was later recognized as being typical for Halley's Comet. At its very close approach to Earth on 24 April of just 15 million kilometres, the comet must have reached a magnitude of between −2 and −4.

On 4 or 5 February 1106, a comet appeared that could immediately be seen in the daytime sky. It was later in the evening sky, and its tail reached a length of 60°. European chroniclers reported that on 16 February, a meteor separated from the comet and fell to Earth. Many authors view this comet as being identical with Ikeya-Seki 1965.

In 1145, Comet Halley was initially in the morning sky, changing to the evening sky after solar conjunction in the middle of May. It was observed between 15 April and 6 July. This return at the time of the Second Crusade is probably depicted in the *Eadwine Psalter*, now at Trinity College, Cambridge.

At Comet Halley's next return in 1222, it was discovered on 3 September by Korean observers, shortly before solar conjunction. According to their report, it was visible in the daytime sky on 9 September. When in the evening sky, it had a tail reaching a length of up to 45°, and a brightness of about 1 mag. Allegedly, when the comet appeared, Genghis Khan had many men put to death in Herat in Afghanistan, although the legend probably related to the conquest of Bamian in 1220, and thus significantly before the comet appeared.

The comet of 1264 remains to this day one of the greatest cometary phenomena in history. It was visible for four months. After its discovery in the evening sky after perihelion, its tail reached a length of 100° in July. The French monk Aegidius de Lessines determined the position of a comet for the first time, which later allowed the orbit to be established. However, contemporaries largely took an astrological view of the comet.

Comet Halley was again in the sky in September 1301. Its perihelion occurred on 25 October. From Central Europe it was circumpolar from 20 September, becoming brighter at the same time. The comet reached its maximum brightness at around 1 mag on 24 September. Petrus Lacepiera (Peter of Limoges) determined its position between 30 September and 6 October. This return has become well known through the famous fresco by Giotto di Bondone in the Cappella degli Scrovegni (the Scrovegni Chapel) in Padua, executed from 1302 to 1305, which shows it as the Star of Bethlehem. The German scholar Conrad of Megenberg commented on its meaning "The comet means much strife and treachery and unfaithfulness and the death of a number of princes and general bloodshed" (based on the Tamman/Veron translation). In contrast to the great attention given it at the beginning of the century, in 1378, Comet Halley was hardly observed at all from Europe, even though it followed a northern path, favourable for observation. Bad weather is probably the reason for the chroniclers' silence.

The Great Comet of 1402, however, caused a great impression, and which was visible from June to September. It was visible in the daytime sky for a week – a record that holds to today. Jacobus Angelus (Jakob Engelhart), in his *Tractatus de Cometis*, described the tail as being 'in the shape of a pyramid'. A Byzantine chronicle, in contrast, spoke of an appearance that looked 'like a spear' on the horizon.

The final reappearance of Comet Halley during this period was in 1456. Chinese astronomers discovered the comet as early as 27 May, probably as a result of an outburst. The position determinations by Paolo Toscanelli have proved to be valuable, because he made observations nearly every day between 8 June and 8 July. In Austria, Georg von Peuerbach in Vienna and Thomas Ebendorfer in Haselbach observed a tail 10° in length. Peuerbach also attempted to measure the parallax of the comet, and determined a distance of at least 1000 German miles (a German mile, or Meile, was about 7.5 kilometres). The efforts of Toscanelli and Peuerbach – although both still believed in the astrological meaning of comets – marked the beginning of scientific research about comets, and the end of the way of looking at comets as objects of astonishment and mythology.

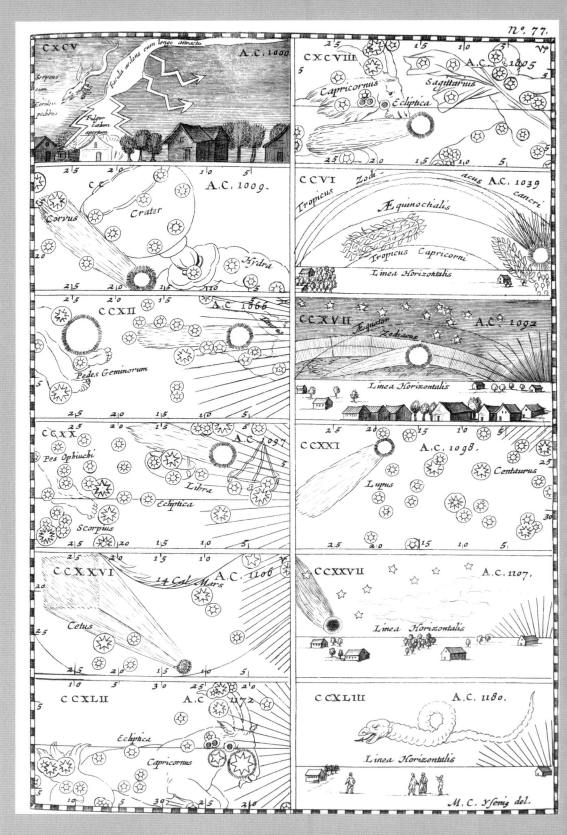

Comets in the Middle Ages from 1000. From the *Theatrum Cometicum* of *Stanislaus Lubinietzky*.

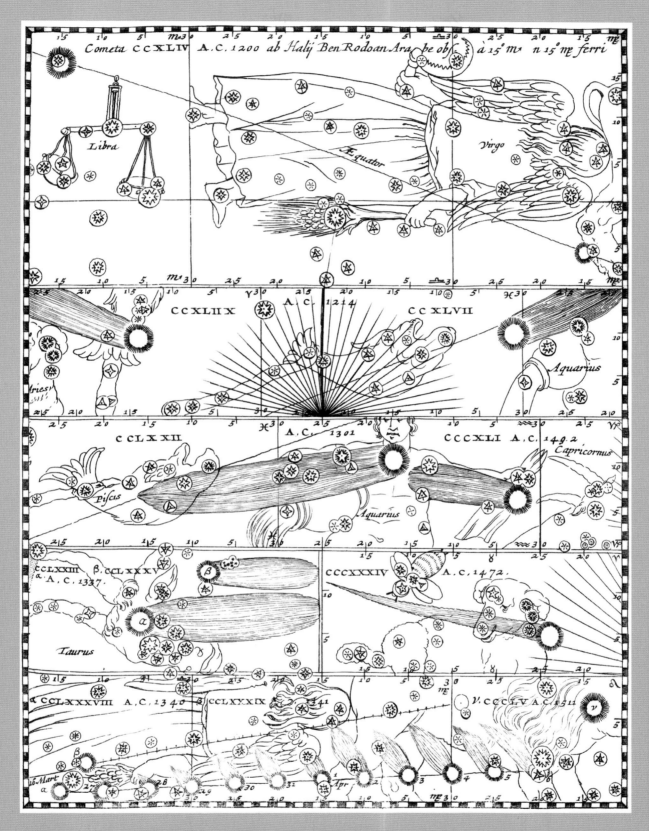

Comets in the Middle Ages from 1200. From the *Theatrum Cometicum* of *Stanislaus Lubinietzky.*

Great Comets

The 30 greatest comets of modern times

⚹ Great Comet of 1471

To this day, no brighter comet has come so close to the Earth. Just 10 million kilometres separated mankind, at the transition from the Middle Ages to modern times, from one of the brightest comets ever to appear. But the old world-view remained firmly established. Many astronomers firmly believed that comets were earthly phenomena, 'below the Moon'.

Data	
Number:	1
Designation:	C/1471 Y1
Old designation:	1471 I
Discovery date:	25 Dec 1471
Discoverer:	Unknown
Perihelion date:	1 Mar 1472
Perihelion distance:	0.4859 AU
Closest Earth approach:	22 Jan 1472
Minimum Earth distance:	0.0696 AU
Maximum magnitude:	−3
Maximum tail length:	50°
Longitude of perihelion:	245.7°
Longitude of ascending node:	292.9°
Orbital inclination:	170.9°
Eccentricity:	1.0

Orbit and visibility

C/1471 Y1 is unique among the bright comets of modern times, because no brighter comet has come so close to the Earth. On 22 January 1472, the separation was just 10 million kilometres. Since then, only the Great Comet of 1556 and Hyakutake in 1996 reached a somewhat comparable value, but neither, however, came so close to our home planet.

The period of visibility was restricted to the time before perihelion on 1 March 1472. The comet first appeared in the constellation of Virgo in the morning sky at the end of December 1471. From there it moved on a generally northerly path into Boötes, passing Arcturus, and into Coma Berenices.

At the end of January 1472, the comet reached Ursa Major, and was visible for the whole night from Central Europe. On the 22nd it passed the North Celestial Pole at a distance of 15°. Closest approach to Earth occurred the following day. At this time the comet was moving faster than 1° per hour and within 24 hours had passed from the spring con-

A portion of the Latin version of the *Nuremberg Chronicle* with the description of the Comet of 1471. *J. Schönsperger*

stellations to the autumn ones. However, the Full Moon on 27 January affected its visibility.

At the end of January, the comet moved back towards the south. It was now an object in the evening sky and was passing through the autumn constellations. There, it disappeared in Cetus around the time of perihelion. Passage past the Sun followed shortly after on 11 March. No observations were recorded after perihelion.

Discovery and observations

It is no longer possible to say, with any certainty, whether the comet was first seen in China or in Europe. Individual sightings had been made as early as Christmas 1471. However, in general, it was first seen in the second week of January 1472. At this time, Korean astronomers gave a tail-length of 4° to 5°. By the third week of January it had increased dramatically to 30°. According to contemporary reports at this time the head of the comet was already 'as large as the Moon'!

The Italian scholar Paolo Toscanelli followed the comet from the 8 to 31 January 1472. In doing so, he noticed the rapidly increasing and, after the 23rd, decreasing, velocity across the sky. Johannes Müller, better known as Regiomontanus, observed from Nuremberg from 13 January to the end of February. He noticed that the tail was always directed in the direction of Gemini – which actually corresponded approximately to the opposite direction to the Sun, which at that time was in Sagittarius. However, this relationship was not obvious to him. Regiomontanus believed that the tail had a daily rotation, because, to him, the comet was not a heavenly body, but was of earthly origin.

On 20 January, Regiomontanus recorded a tail length of about 50°. The comet must have presented a magnificent sight at this time. Chinese sources even claim that, when close to the Earth, it was visible in daylight. However, that is believed to be rather unlikely, because perihelion only came about a month later. With its magnitude of –3, however, C/1471 Y1 was one of the brightest comets that has ever been observed outside twilight.

Background and public reaction

Because of the uncertainty of historical sources, it is difficult to estimate, nowadays, how bright the Great Comet of 1471 became. The most reliable observations suggest a bright, but not an exceptionally bright, phenomenon. Accordingly, the comet would probably not have been visible in daylight.

Eberhard Schleusinger from Bamberg took the Great Comet as an opportunity to carry out a distance determination. He did this, starting from the Earth's circumference as being 913 German miles (Meilen), and that the distance of the Moon was 33 times this value. He determined the distance of the comet from a daily observation of its motion relative to the star Spica. From a value of 6°, he calculated the distance as 8200 Meilen (equivalent to about 62 000 km) – which confirmed the Aristotelean view that comets were not objects within the outer fiery spheres, but instead belonged to the sublunar region influenced by the Earth. Regiomontanus, who became famous through his tables of ephemerides for seamen, accepted Schleusinger's calculation. It was published after his death and has thus falsely been ascribed to him by many different sources..

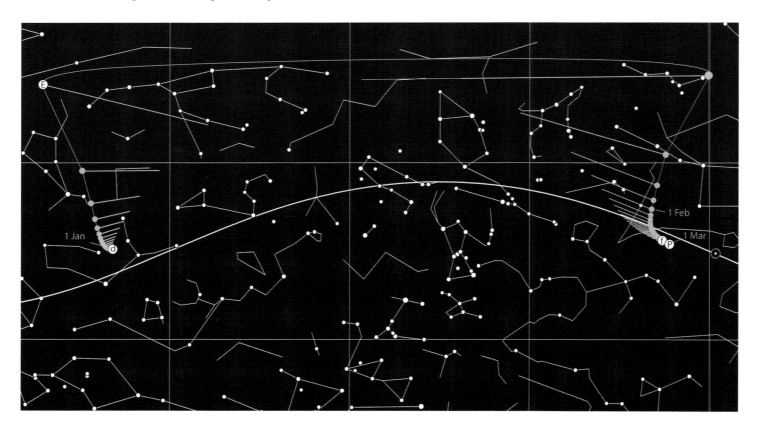

☄ Comet Halley 1531

The Reformation and book printing were reshaping Europe when Halley's Comet made its first appearance in modern times in 1531. While Protestant clergymen propagated the idea that the comet was a divine sign to Christianity about changing from old views to new ones, in Ingolstadt Peter Apian observed that the tail always pointed directly away from the Sun.

Data	
Number:	2
Designation:	1P/1531 P1 Halley
Old designation:	01531 I
Discovery date:	5 Aug 1531
Discoverer:	Unknown
Perihelion date:	26 Aug 1531
Perihelion distance:	0.5812 AU
Closest Earth approach:	14 Aug 1531
Minimum Earth distance:	0.4414 AU
Maximum magnitude:	0
Maximum tail length:	15°
Longitude of perihelion:	107.0°
Longitude of ascending node:	52.1°
Orbital inclination:	162.9°
Eccentricity:	0.9674895

Orbit and visibility

The return of Halley's Comet in 1531 was favourable for northern-hemisphere observers. The comet passed north of the Sun at closest Earth approach and was thus well-positioned for European viewers. As such, the period of best visibility was before perihelion.

At the end of July, when the first sightings of the comet were made, it was in the morning sky east of Boötes. Its path took it towards the north, though an area that later became the constellation Lynx, below the paws of the Great Bear, Ursa Major. On 8 August it reached its greatest northern declination there. In Central Europe it was thus circumpolar, and could therefore be seen throughout the night, with the best viewing conditions after evening twilight and before dawn.

In the middle of August, the comet passed the Sun, reaching the minimum distance of 23° on the 16th. The comet was then north of Leo. Visibility was now restricted to the evening sky. The distance from the Sun increased to 36° on 29 August. Perihelion occurred shortly before that date.

In September the comet again appeared to move towards the Sun. It passed southwards through Virgo, where the last observations were made.

Discovery and observations

The comet was first sighted from China on 5 August, when it was between Auriga and Gemini. At that time it already had a tail that was about 2° long.

The most influential European observer was Peter Apian, who followed the comet from Ingolstadt. He noted its position between 13 and 23 August.

By 13 August the tail had developed to an imposing length of more than 13°. It stretched towards the north, such that in Central Europe it could be seen around midnight above the northern horizon.

The greatest magnitude came on 18 August. According to modern calculations the comet must have reached between magnitude 0 and 1. From Europe the last observation was on 3 September. In the Far East, because of the more southerly locations, it could be followed until the 8th.

Background and public reaction

Halley's Comet was the first of three bright comets in just two years. With its magnitude of up to −1, C/1532 R1 must have been significantly brighter. It followed a similar path to Halley's, from Gemini, across Leo, and into Virgo and, overall, was visible for 119 days between 2 September and 26 December 1532. Tail lengths up to 15° were reported. C/1533 M1 appeared in the sky in July and August 1533 for about 80 days. Its magnitude amounted to about 0. The first of Halley's Comet's returns in modern times coincided with the rise in printing and also the Reformation. In Germany, where both started, both pamphlets and broadsheets first appeared, describing the occurrence and its significance. In particular, Protestant clergy reinterpreted the traditional astrological explanation into Christian terms, in which the comet, as 'God's scourge', announced a turning point in history,

and which should cause the faithful to adopt the true faith through repentance and reform. Martin Luther also saw the comet of 1531 as a sign from God and, in a letter of 18 August, remarked that the comet "did not mean anything good".

But the old astrological doctrines flourished. The Swiss reformer, Huldrych Zwingli himself, saw the comet as a prediction of his early death, which actually occurred on 11 November. An astrological comet pamphlet by Paracelsus, the physician, about the negative consequences of the comet, which was, of all things, dedicated to Zwingli, may well have had an influence.

Pioneering astronomical work was carried out by Apian (Peter Bienewitz) of Ingolstadt. He followed the comet's path across the sky and determined the position of the comet over many weeks. Admittedly, his position determinations were fairly inaccurate, but he noticed that the comet's tail was always directed away from the Sun. Apian confirmed this relationship, first found in Halley's Comet of 1531, with subsequent comets in 1532, 1533, 1538 and 1539. He published his results in his monumental work *Astronomicum Caesaraeum*. Even then, Apian also had the idea of determining the distance of the comet, by means of the parallax, that is, the apparent shift in its angle as seen from different places on Earth. With the means at his disposal, however, he was unable to carry out this procedure. For Halley's Comet of 1531 he gave a straight-line path rather than a curved orbit, as was considered to apply to comets until well into the seventeenth century.

Title page of a 1531 treatise on comets. *J. Schöner*

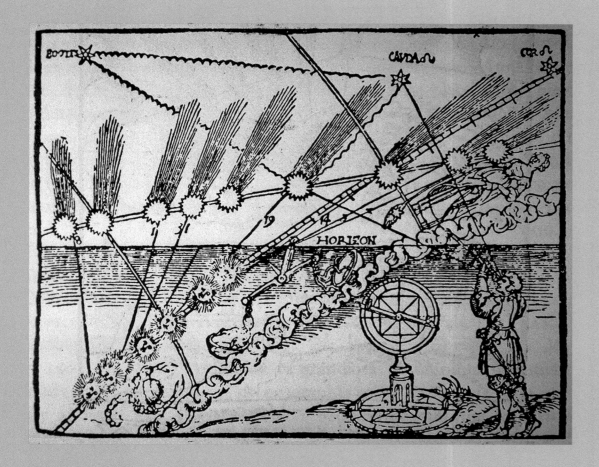

◄ By the time of Halley's return in 1531, the relationship between the position of the Sun and the direction of the tail had been established in the western world. This illustration shows measuring the position of the comet relative to stars. Woodcut. *P. Apian*

▼ Illustration of the bright comet C/1532 R1, which appeared just over a year after Halley's Comet. Watercolour. *Unknown artist*

☄ Great Comet of 1556

wo frightful earthquakes marked the year 1556. In China almost one million people were killed and, in the Middle East, wide areas of the city of Constantinople (present day Istanbul) were destroyed. To contemporary witnesses the event responsible was definite: the Great Comet that appeared in the early part of 1556. Like Hyakutake 430 years later, its path took it close to the Pole Star.

The Comet of 1556 was regarded as the cause of the great earthquake in Constantinople (Istanbul) on 10 May 1556. Coloured wood-cut. *Hermann Gall*

Orbit and visibility

The path of the Comet of 1556 had certain similarities to that of Comet Hyakutake. Its track across the sky took it, like the latter, past the North Celestial Pole. Closest approach to Earth likewise came more

than a month before perihelion. C/1556 D1, however, approached Earth some 3 million kilometres closer than Hyakutake's distance of 15 million kilometres.

At discovery, the comet was west of Spica in Virgo. As such it was almost opposite the Sun and was thus visible throughout the night. At

Data	
Number:	3
Designation:	C/1556 D1
Old designation:	1556 I
Discovery date:	27 Feb 1556
Discoverer:	Joachim Heller
Perihelion date:	22 Apr 1556
Perihelion distance:	0.49 AU
Closest Earth approach:	12 Mar 1556
Minimum Earth distance:	0.0834 AU
Maximum magnitude:	−2
Maximum tail length:	10°
Longitude of perihelion:	100.9°
Longitude of ascending node:	181.4°
Orbital inclination:	32.4°
Eccentricity:	1.0

the beginning of March, the comet rapidly moved towards the north through Virgo, across Coma Berenices and into what later became the constellation of Canes Venatici. On 13 March, the comet crossed the 'tail' of Ursa Major, and this was the day it came closest to Earth.

In the following days, the comet passed through the northern region of the sky, across Ursa Minor and Draco. The circumpolar location improved visibility conditions for European latitudes. The comet's tail pointed towards the southwest. This resulted in the rare effect that, during the first half of the night, the tail stood up above the horizon. At the end of March, the comet moved through the constellations of Cepheus and Cassiopeia. In doing so, within a few days, it moved from the spring constellations to those of autumn.

Observational conditions rapidly worsened. The comet was now only visible in the morning twilight. On 1 April, the comet was in the constellation of Andromeda, rapidly moving south. Perihelion took place on 22 April, as the comet was passing through the northern portion of Pisces. The last observations were made in Aries, when the comet was about 30° away from the Sun.

Discovery and observations

The Nuremberg calendar maker, Joachim Heller, was on a journey in the Fichtel Mountains in Bavaria near Bad Berneck, when, on 27 February 1556, he saw a 'great fiery unusual star' beneath Spica in Virgo. He was unable to see any tail.

On 3 March, he saw the comet again after his return to Nuremberg and learned that it had been noticed from the town on previous days. By this time a small tail was visible. From the beginning of March, the comet began to cause attention in other locations. In China it was seen from 1 March, and in Mexico from 6 March.

Heller followed the comet subsequently from 6 March to 22 April. Paul Fabricius, the personal physician to the emperor, Charles V, and who was also an astronomer in Vienna, carried out nearly daily observations from 5 March. A contemporary observer likened the appearance in the sky to the 'flame of a candle, blown by the wind'. In the middle of March, Fabricius compared the brightness to first-magnitude stars, and established that the comet appeared brighter. Contrary to this, the Belgian astronomer Cornelius Gemma likened the brightness to that of Jupiter, and its colour to that of Mars. Nowadays, a maximum magnitude of −2 is assumed.

However, the tail does not seem to have increased in length. If we take the sketches by Gemma on a chart as a basis, it did not ex-

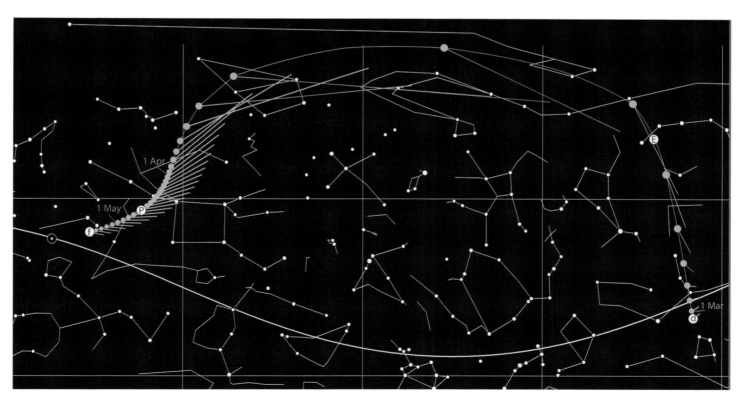

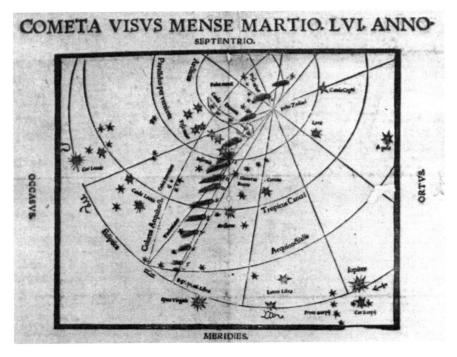

COMETA VISVS MENSE MARTIO. LVI. ANNO
SEPTENTRIO.

MERIDIES.

◄ Path of the Comet (*top*) and view over Nuremburg (*centre*). *D. Fabricius (top), H. Weygel (centre)*

ceed 5 to 10 degrees. Strangely enough, at closest approach to Earth on 13 March, the comet was briefly lost by many observers. This could be ascribed to the weather, but may perhaps be because of the activity of the comet itself.

From Europe, the comet was seen for the last time in the middle of April. Chinese reports continue to report it until 10 May.

Background and public reaction

Contemporary witnesses linked the Great Comet of 1556 with two disastrous earthquakes in the same year. However, these occurred both before and after the comet was visible. The earthquake of 23 January 1556 in the Chinese province of Shaanxi remains to this day the most deadly earthquake of all time. Some 830 000 people were killed. There were significantly fewer victims of the earthquake on 10 May in Constantinople, but large portions of the city were destroyed. The minarets of the converted Hagia Sophia mosque collapsed as well as large sections of the city wall and its towers.

According to legend, the Great Comet caused Charles V, the Spanish king and Holy Roman Emperor to abdicate and enter a monastery. As a result, the comet was also known as 'Charles' Comet'.

Some 300 years after the appearance of the comet, it once again made the headlines. This was due to calculations by the English astronomer John Russell Hind, who considered that the Comet of 1556 was a return of the Great Comet of 1264. He predicted a further return for 1850, but astronomers waited for this in vain. According to current knowledge, no link between the comets of 1264 and 1556 can be established.

◄ A contemporary illustration of the Comet of 1556 from southern Germany.

► Broadsheet about the comet, which links the celestial body with the earthquake in Constantinople. *H. Gall*

Ein erschröcklich wunderzeichen / von zweyen Erdbidemen /
welche geschehen seind zu Rossanna vnnd Constantinopel.
Im M. D. LVj. Jar.

Neben dem Cometen / so den fünfften tag Martij allhie zu Constantinopel gesehen ist worden / mit ainem seer langen schwantz / welcher bey zwölff tagen geschinen / waiß ich euch glaubwürdig anzuzaigen / Das sich in vergangnem Monat Aprilis / in einer Statt Rossanna genant / in Astopia gelegen / ain seer grosser vnd erschrecklicher Erdbidem erhaben hat / der fast durch die gantze Statt / vñ darin vil menschen verdorben vnd abkommen seind / hat auch von gemelter Statt auß gewehret / fast in die dreissig oder viertzig meyl wegs. Darnach den x. tag diß monats Maij / hat sich gleich sahls allhie zu Constantinopel diser erschrecklicher Erdbidem auch erhaben / vnd in der gantzen Statt mercklichen schaden gethan / dan es hat vil Thürn vñ onzalbare Gebäw eingeworffen / welche auch vil menschen erschlagen haben. Vnser herberg / da wir seind / ist auch ohn schaden nicht dauon komen / also / das wir alle augenblicklich besorgten / wir wurden sampt dem Gebäw zu grund gehen / welches auch nicht weyt gefehlet hat / weyl sich diser vnd andere orth zu einfallen schicktñ. Aber der herrliche Tempel S. Sophie (welchen gemainklich die Türckischen Kayser pflegen zu besuchen) hat durch solchen Erdbidem grossen schaden empfangen. So ist auch das Hadrianopolitanisch thor / vnnd ain grosser tail der Mawrn eingefallen / dauon der Türck trefflich hart erschracken. Diser Erdbidem hat drey tag gewehret / Den dritten hernach / welcher ist der xiij. diß Monats / ist ain Stern erschinen aines wunderbarlichen hällen scheins / vnd ist dem Mond so nahen gestanden / als ob er hart an jne hieng. Das hab ich euch für Newe zeytung auff diß mal schreiben wöllen. Geben in grosser eyl zu Constantinopel / den xv. Maij / M. D. LVj. Jar.

Dises Wunderzaichen bringt on zweifel mit sich / das der Jüngste tag vor der thür ist / auff welchen Gott richten will den krayß des Erdbodens / mit Gerechtigkait / durch ainen Man / in welchen ers beschlossen hat / Nämlich durch Jesum Christum / welchem er alles gericht vbergeben hat / das er die so an sein wort glauben / soll aufferwecken zum ewigen leben / vnd selig machen / Die vnglaubigen aber / die jn verlestert / vnnd sein wort verfolgt haben / soll aufferwecken zum gericht / vnd ewigklich verdamen: Denn das gilt dem Türcken so wol / als den verachtern Göttlichs worts: Das der Tempel Sophie (welcher zuuor durch den Bapst eingeweyhet ist) mit grossem schrecken vnd zittern ist eingefallen. Nun haißt Sophia auff teutsch Weißhait / vnd darumb / das sy Gottes weyßhait Jesum Christum auß dem Tempel vertriben / vnd Gottes wort nicht leydñ künden / sien an desselben statt jre aygne weyßhait / lügen / Abgötterey vil verfürung / vnd wöllen noch zu kainer Buß komen / so will sy Gott zu gericht fordern / das sy mit erschrecklichem zittern vnd zagen sehen sollen / wie der Prophet sagt: In welchen sy gestochen / das ist / was für ainen Herren vnd Christum sy veracht haben. Wir Christen aber sollen wachen vnd bätten / nüchtern sein / vnd vns auff disen frölichen tag gerüst machen mit Gottes wort / vnd rechtem glauben / auff das wir mit den Gotlosen verächtern kainen thail haben / Sonder das wir würdig werden allem disem vnglück zu entpfliehen / Vnd zustehen mit gutem gewissen für des menschen Son / als die wir im glauben durch sine wunden gehaylet / in seinem Blut gerecht / vnd in disem ainigen Christo ewigklich selig werden / Amen.

zu Nürmberg bey Herman Gall / Brieffmaler / in der Kotgassen.

☄ Great Comet of 1577

A comet 'as bright as the Moon' appeared in the evening sky in November 1577. The general public, intimidated by numerous broadsheets, trembled at the sight of this 'horrible' star. Meanwhile the astronomers – who also acted as astrologers – fell out amongst themselves. Some saw an ominous portent in the red colour of the tail, while others attempted to determine the distance. The fight between superstition and science was to dominate the discussion about the meaning of comets for some time to come.

The Great Comet of 1577 over Prague. Woodcut. *G. J. von Datschitz*

Data	
Number:	4
Designation:	C/1577 V1
Old designation:	1577 I
Discovery date:	1 Nov 1577
Discoverer:	Unknown
Perihelion date:	27 Oct 1577
Perihelion distance:	0.1775 AU
Closest Earth approach:	10 Nov 1577
Minimum Earth distance:	0.6271 AU
Maximum magnitude:	–7
Maximum tail length:	30°
Longitude of perihelion:	255.7°
Longitude of ascending node:	31.2°
Orbital inclination:	104.9°
Eccentricity:	1.0

All dates are in the Julian calendar.

Orbit and visibility

When the comet was discovered, its perihelion on 27 October 1577 had already passed by a few days. It should have been easily visible in the evening sky in October in Boötes, but no sightings have been handed down from that period.

In the first days of November, the comet was in Scorpius, about 15° from the Sun in the evening twilight. The tail pointed towards the south, so that initially, for a European observer, there was the unusual sight of a comet whose tail sloped down towards the horizon.

On 5 November, the comet passed the bright star Antares. At closest approach to Earth on 11 November it was already in the constellation

of Sagittarius. The time for observations after sunset had increased by three hours, and the comet could be seen clearly under dark skies. The tail now turned into a more easterly direction, so that from Central Europe, when the comet was in Aquila, it appeared to run more-or-less parallel to the horizon.

The comet then rapidly increased its distance from Earth. On 25 November, the Full Moon interfered with its detection. On 1 December the comet entered Equuleus, and at the turn of the year, 1577–78 was in Pegasus. It was there until the last sightings took place in January 1578.

Discovery and observations

The first observations of the Great Comet of 1577 came from Peru, which had been conquered by Francisco Pizarro barely 45 years before. Here, the comet was first noticed on 1 November, when 'it shone through the clouds like the Moon'. In Japan, where it was also detected, on 8 November, the comet 'appeared as bright as the Moon'. According to modern estimates the magnitude at that period was as bright as –7 magnitudes!

In Europe, the comet was seen in the middle of November. Bartholomäus Scultetus in Görlitz saw the comet on 10 November as 'great red mass, spewing fire, with the end terminating in smoke'. Two days later, Michael Mästlin in Tübingen noted the enormous tail, about 30° long. Tycho Brahe, at a disadvantage because of his northern location on the Danish island of Hven, accidentally discovered the comet in the evening twilight, in Pisces, on 13 November. It appeared to him 'as bright as Venus when closest to the Earth' and at the beginning of darkness found the tail to have a length of 22° and a width of 2.5°. Brahe described the head as 'golden' with 'red rays'. He carefully followed the comet over the following days. In doing so, he noticed the decrease in its apparent motion in the sky from approxi-

Newe Zeytung von dem Cometen/

So jetzt im Nouember dises 1577. Jars erschinen/ vnd
beschreibung der bedeütung desselbigen.

Als man zalt 1577. Jar im Nouember/erscheine ein

Comet/welches gleichen zü vnsern zepten im Jar 1531. vnd 1556. gesehen worden. Ist aber den 13. vnd 14. tag gemelten Monats im Stainbock nicht weyt vom Saturno vnd Mon
gegen Mitternacht gestanden/vnnd seinen Schwantz gegen dem auffgang der Sonnen geworffen. Solch
Hirnlisch zaichen hat wol sein natürlich vrsachen/so auß den Finsternussen/welcher zwo diß Jar geschehen
seind/vnd von dem Fewrigen schwefelichen dämpffen/so auß krafft der herrschenden Planeten/als Martis
vnd Saturni auffgehoben worden/genommen werden/hat aber darneben auch sein schreckliche bedeüt-
tung/welche nach lehr vnd erfarnuß der alten Astrologen vnnd Naturkündiger allhie erzelet sollen werden.
Dann erstlich bedeüt diß Prodigium ein grosse Hitz/welcher halben das Erdtrich soll außgetrucknet vnnd
vnfruchtbar gemacht werden. Zum andern/ein gemein Sterben von wegen der Lufft/auß welcher das Hu-
midum radicale/das ist/natürliche feüchtigkeit/wirdt gleichsam auffgesogen/oder ja vergifft von warmmen
groben trüben dünsten. Zum dritten/grewliche Krieg vnd Auffrühr/sampt verenderung Gaistlicher vnd
Weltlicher satzungen vnd Policeyen. Zum vierdten/abgang grosser Herren/Fürsten vnd Monarchen. Zum
fünfften/schädliche außlauff der Flüß vnd Meers/Item Sturmwind vnd Erdbidem. Zum sechsten ist diß
Prodigium ein zaichen/dardurch vns Gott vermanet (dieweil sein straff nicht weyt) ein Gottselig leben
zufüren/vnd mit emsigem Gebett bey jm anzuhalten/die wolnerdinte straff von vuns abzuwenden. Dann
waun Gott straffen will/so pflegt Er allwegen durch solche Zaichen seinen zorn anzuzeigen/wie dann auß
allen Historien wol zusehen. Es erstreckt sich aber die bedeütung dises Cometens biß auff zwölff Jar. Der
Allmechtige barmhertzige Gott wöll vns sein genad verlethen/das wir sollich Zornzaichen recht bedencken/
vnd zü einem Gnadenzaichen verkeren mögen/Amen.

¶ Getruckt zü Augspurg/durch Valentin Schönigk/auff vnser Frawen Thor/
vnd bey Hanns Schultes Buetzmaler vnd Formschneyder zufinden.

Two broadsheets about the Great Comet of 1577 with views of the city of Augsburg. Both representations relate to 12 November 1577. *Valentin Schönigk (above), Bartholme Käppeler (opposite)*

Ein kurtze erinnerung / von dem Cometen / so auff den
12. tag Nouembris des 1577. Jars zu Augspurg erschinen /
vnd erstmals gesehen worden.

Ach dem dañ abermals mit disem new schwebenden Liecht vnd Cometen / Got seinen zorn vnd straff wider das Menschlich geschlecht / vmb seiner vilfeltigen Sünden willen offenbaret vnd zuerkennen gibt: ist vnleugbar / das er dasselbig mit allerley straffen vnd plagen werde heimsuchen / dann auch Lucanus sagt: Et cælo nunquam spectatum impunè Cometem, das ist: Es ist kein Comet nie gesehen worden / darauff nit besundere straffen erfolget weren. Will derwegen dem gutherzigen Leser auffs kürzest für die Augen stellen / was diser new vnd noch schwebende Comet / auß Astrologischem grund für bedeutung vnd Göttlicher Ruthen anzeigung habe / hindan gesetzt alles anders / vnd allein auffs end gehn / daran dem gemeinen Mann am aller meisten gelegen. Dieweil dañ diser Comet / (wie die Gelehrten wissen) ganz vnd gar Saturnischer art vnd eygenschafft / so drewet er erstlich vilen Leüten Melancholische kranckheiten / Kopffwehe / Schwindsucht / viertäglich Fieber / fallenden Wehtagen / den Kreps / Auffsatz / böse Rauden / Blutflüß / den Schlag / in Summa langwirige Kranckheiten. Darnach drewet er Krieg zwischen Potentaten vnd jren Vnderthonen / zu dem auch viler jammer vnd vnglück / anderstwo anck / zwitracht / belegerung der Strassen / Gifft / grosser Fürsten vnd Herren absterben / verachtung Gaystlicher Personen / vnd verfolgung der Gotseligen.

Zum dritten / ist zubesorgen / wir werden einen rauhen / von kälte vnd Schnee Winter haben / darauff dann auch ein groß Gewesser anfallen / vnnd an vilen orten den Feldbaw verderben mag. Etlich sezen darzu Hunger / Pestilenz / vnnd Auffruhr / welches sie durch den Cometea / so vnder der Regierung Kaysers Hainrichen des vierdten im 1097. Jar erschinen / bestettigen. Dann da ist erstlich grosse theüre zeit vnd Pestilenz erfolgt. Nachmals hat der hochberümbt Herzog Gottfrid Billionius die Statt Hierusalem mit Heeres krafft erobert vnd eingenommen.

Zum vierdten wird auch angedeut / das schand vnd laster / vnd vnzucht vber hand nemen / was dann auß solchem wilden wüsten vnd vnzüchtigem leben erfolge / kan ein jeder selbst ermessen. Diß alles möcht am meisten die berüren / so vnter dem jrdischen Triangel / arüon besich anderer Schrifften / die ich vmb kürze willen hie vnterlaß / Sonderlich aber / was dem Capricorno vnterworffen / wer aber meint / er werde diß vnglück gefreyet vnd vberhaben sein / der hinderdenck was die Finsternus im 1572. Jar / vnter dem Himlischen Zaichen des Steinbocks ergangen / hie in disen vnnd anderen der gleichen orten hab angericht / der wird freylich befinden / das die zeit nit on sunder wehklagen der Leut sey fürüber gangen / Wie vil mehr vngemachs haben wir von disem zorn zeychen zugewarten.

Jesu Christe / ein Sun des waren lebendigen Gottes / Regier vns durch dein hayligen Geist / das wir durch dein Wort / Zaychen vnnd geschöpff vermanet / dich wahren Gott / sampt dem Vatter vnd hayligen Geist recht erkennen vnd anrüffen / Amen.

Zu Augspurg bey Bartholme Käppeler Brieffmaler / im kleinen Sachsen Geßlein.

mately 3.5° to 1.5° per day on 30 November and 25'
per day on 15 January.

The extremely bright tail started to fade at the beginning of December. On 28 November, Cornelius Gemma reported a doubled tail. Finally, on Christmas Eve, because of the bright Moon, the tail was hardly visible. When Tycho Brahe saw the comet for the last time on 26 January, it was barely visible to the naked eye.

Background and public reaction

C/1557 V1 was one of the intrinsically brightest comets ever. It passed the Sun at the relatively close distance of 0.18 astronomical units, but came no closer to the Earth than 90 million kilometres.

The Great Comet of 1577 caused the earliest peak of comet hysteria in Europe with about 200 broadsheets and texts, most of which were devoted to the theme of the negative consequences of its apparition. The comet was, however, the first in history to be carefully followed scientifically. Credit for this goes to the Danish astronomer, Tycho Brahe. From his determinations of position, he believed that the comet arose from the rays of the Sun – and actually its path across the sky after perihelion appeared to confirm this. On 23 November, Tycho tried to determine the parallax of the comet when compared with the star ε Pegasi. Thaddaeus Hagecius (Tadeáš Hájek) in Prague also observed the positions of the comet and star on the same day. However, no parallax was detectable, whereupon Tycho assumed a minimum distance of 230 Earth radii. He gave the tail's length at 70 000 German miles, or about three times the Moon's distance.

Tycho, who rejected the Copernican world view and favoured a mixed system with heliocentric and geocentric elements, saw comets as bodies orbiting the Sun and not – as widely accepted at that time – as vapours emitted by the Earth. In doing so, he took a definite stance in the discussion of whether comets were sublunar or supralunar phenomena. Hagecius, for example, accepted the classical Aristotelian interpretation 'beneath the Moon'.

Tycho, however, also considered the astrological significance of comets, and predicted major disasters for Spain – which he derived from the southern location of the comet in the sky and its appearance in the eighth house, which was associated with death. In addition the red colour of the comet was a bad sign. Legend also has it that the court astrologers to Elizabeth I of England

▲ Coloured print of the Comet of 1577 over Nuremberg – a view towards the south across the castle and city. Detail. *Georg Mack*

▲ A Turkish astronomer observing the Comet of 1577 over Constantinople (Istanbul). He uses a quadrant for determining positions. *Mustafa Ali*

were also alarmed by the effects of the comet. They advised her against viewing the comet, because it would bring bad luck. The queen, however, rejected the advice and, on looking at the comet, remarked laconically "Iacta est alia." ('The die is cast.') She reigned for another 25 years.

☄ Comet Halley 1607

I t was beyond Kepler's powers of imagination to believe that the comet of 1607 would return in 1682. The 'discoverer' of Keplerian orbits believed that comets originated from the 'ether' near the Earth, and would move perpetually on straight paths out of the Solar System – even though he had himself observed the curved path of the comet of 1607 around the Sun. It was not until 100 years later that the time was ripe for the comet of 1607 to be ascribed an orbit in accordance with Kepler's own laws.

Data	
Number:	5
Designation:	P1/Halley
Old designation:	1607 I
Discovery date:	20 Sep 1607
Discoverer:	Unknown
Perihelion date:	27 Oct 1607
Perihelion distance:	0.5836 AU
Closest Earth approach:	29 Sep 1607
Minimum Earth distance:	0.2450 AU
Maximum magnitude:	−1
Maximum tail length:	10°
Longitude of perihelion:	107.5°
Longitude of ascending node:	53.0°
Orbital inclination:	162.9°
Eccentricity:	0.9674895

All dates are in the Gregorian calendar.

Orbit and visibility

At Halley's Comet's return in 1607, observers in the Earth's northern hemisphere were privileged. The period of visibility before perihelion found the comet at high northern declinations, so observers in Europe were allowed a good view of the comet.

At discovery at the end of September 1607, Halley's Comet was on the border of the constellations of Gemini and Lynx in the morning sky. From the 20 September, it was circumpolar for observers in Central Europe and thus observable throughout the night, without interference from twilight. It crossed the southern portions of Ursa Major and Canes Venatici at the end of September and beginning of October, and finally moved towards the south. In doing so it passed close to and north of Arcturus.

In October the observational window had shrunk to the evening sky. In the middle of the month, the comet moved south through Ser-

pens. In making for the Sun, it became ever more difficult to see. It finally reached the Sun on 21 November in Scorpius. It had already passed perihelion on 27 October. After conjunction with the Sun, the comet remained unobservable, because by then it was too far away from the Earth.

Discovery and observations

Chinese astronomers were the first people to see the comet. On 21 September, they noted a 'pale broom star' with a tail 3° long. The magnitude must have been about 3 at the time.

In Europe, the first sightings recorded are from 23 September. Johannes Kepler found the comet by accident on 26 September, when he was standing, watching fireworks from a bridge over the Vltava in Prague. He observed the comet continuously until 26 October, noting down the current positions. Another careful observer was Thomas Harriot in England, whose series of observations covered 1 to 23 October.

According to Harriot, the comet was as bright as Arcturus on 1 October. On 9 October the magnitude had faded to about 2, and to about 2.8 on 23 October. The last observations from Europe were reported on this date. In China, the comet was followed until 12 October. After conjunction with the Sun, no observations are known.

Background and public reaction

Halley's apparition in 1607 was the last before the invention of the telescope – Kepler, Harriot and other astronomers of the time were forced to rely upon the naked eye. The main focus of their observations lay in determining the positions of the comet on the sky, to describe its path.

Kepler correctly obtained a curved orbit from his observations, a sketch of which he also gave in his work about the comet of 1607. But he revised this result in his work of 1619, in which he summarized his cometary observations obtained up to that date. Here, he spent a lot of effort in constructing a path that took the comet away from the Earth in a straight line.

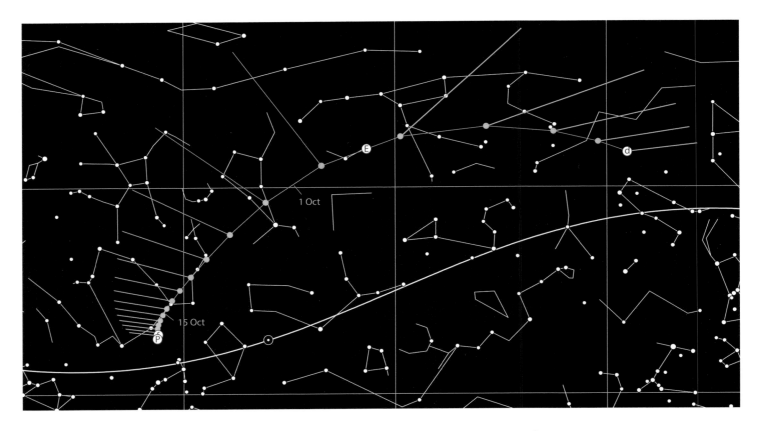

Obviously Kepler was unable to discard the idea of a linear path, even though he had himself advanced epoch-making discoveries about planetary orbits. In his work on the comet of 1607, he repeated the assertions then circulating about comets. According to these, comets formed spontaneously out of the 'ether' of the sublunary sphere. A specific 'spirit' accompanied each comet and died with it. The tails arose from the release of cometary particles through rays from the Sun, like steam or mist. If a comet's tail were to come into contact with the Earth, it would bring contamination and death. In addition, the essay also contained currently popular astrological interpretations – after all, astrology was Kepler's 'bread-and-butter' job.

Halley's Comet of 1607 made no major impression on the general public. Only 17 tracts and broadsheets were published in German. They gave various dates for the appearance of the comet – depending on whether it was observed from a Catholic or Protestant country and thus respectively in the new Gregorian calendar, introduced in 1582, or the old Julian calendar. The difference amounted to ten days. It was only in 1700 that the Protestant states in the Holy Roman Empire adopted the calendar reform. In England and its colonies the old calendar remained in use until 1752.

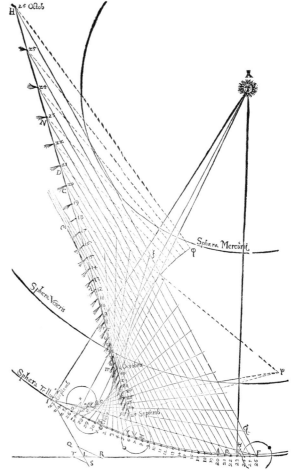

Contrary to his own observations, Kepler constructed a straight path for the comet. From *De Cometis Libellis Tres*, 1619. *J. Kepler*

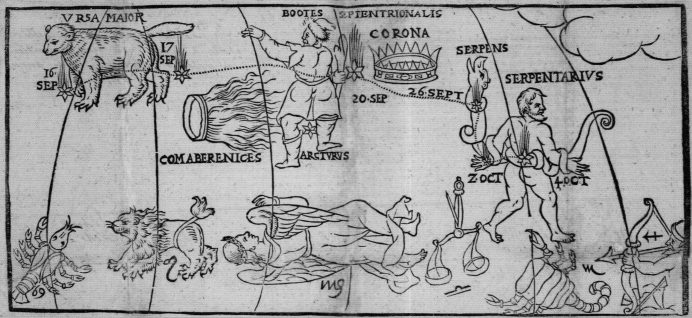

The path of Halley's Comet through the constellations. The rather 'angular' course shown in the upper, reversed representation, is remarkable. The lower, more modern, depiction was prepared sixty years later. *J. D. Herlitz (top), S. Lubinietzky (bottom)*

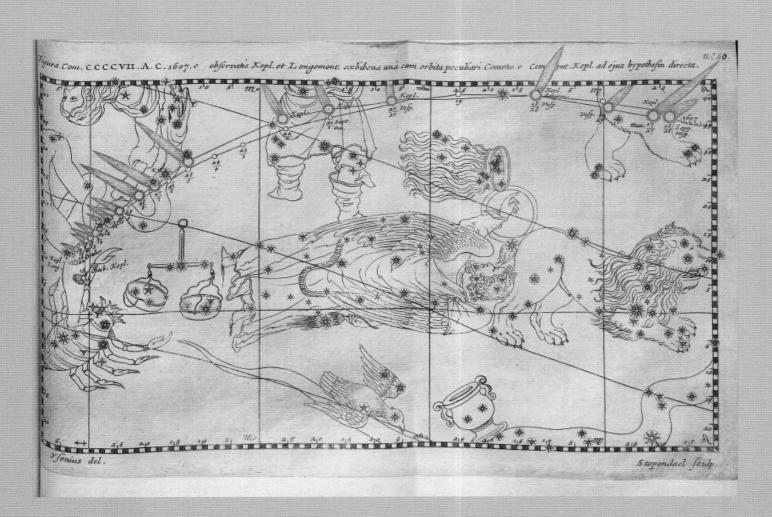

☄ Great Comet of 1618

In the autumn of 1618, the devastating Thirty Years' War had just begun, when three bright comets appeared on the scene at the same time. The general public saw this as an omen for the coming disaster and numerous broadsheets urged believers to reform and repent. Yet at the same time, astronomers such as Johannes Kepler were able, for the first time, to scrutinize the comet with a telescope, and were about to demolish myths that were hundreds of years old.

The Great Comet of 1618. *François Schillemans*

Orbit and visibility

When C/1618 W1, the second bright comet of the year, was discovered, it had already passed perihelion. In the middle of November it was located in the area of Scorpius and Lupus, and was therefore vi-sible from southern locations. For Central Europe it was initially lost in the glare from the Sun. That first changed at the beginning of December, as the comet moved north into the constellation of Libra. An observational window thus opened in the morning sky, which subsequently increased steadily. For one thing, the comet was moving

Number:	6
Designation:	C/1618 W1
Old designation:	1618 II
Discovery date:	16 Nov 1618
Discoverer:	García de Silva y Figueroa
Perihelion date:	8 Nov 1618
Perihelion distance:	0.3895 AU
Closest Earth approach:	6 Dec 1618
Minimum Earth distance:	0.36 AU
Maximum magnitude:	0
Maximum tail length:	100°
Longitude of perihelion:	287.4°
Longitude of ascending node:	81.0°
Orbital inclination:	37.2°
Eccentricity:	1.0

All dates are in the Gregorian calendar.

in a northerly direction, and for another, it also appeared to be moving away from the Sun.

By the beginning of December, it had reached Boötes. This is where its closest approach to Earth occurred on 6 December. On 9 December it passed about 5° to the east of the constellation's brightest star, Arcturus. By this time the comet was visible throughout the second half of the night and had reached its greatest brightness.

In the second half of December, the path had again turned north. By 21 December, the comet crossed the 'tail' of Ursa Major. It thus became circumpolar and was visible throughout the night from Central Europe. The Full Moon interfered with observations on 31 December.

In January 1619, the comet passed north of Ursa Major, and it was here that it was last seen.

At the same time as C/1618 W1, C/1618 V1 was also in the morning sky. Also discovered after perihelion; it moved in a southerly direction from Libra towards Corvus and Hydra. It was therefore more difficult to observe from Europe, and was always closer to the horizon.

Discovery and observations

The first observer of C/1618 W1 could have been the Spanish ambassador in Persia, García de Silva y Figueroa. He noticed the comet on the morning of 23 or 24 November 1618, and compared the head of the comet with Venus. The bright tail appeared to be curved.

At this time, C/1618 V1 had already been seen for about twelve days. That comet had an impressive tail of about 60° in length, which stretched from Libra into Corvus. However, C/1618 V1 was observed only until 29 November in Europe and until 4 December in China.

C/1618 W1 was followed by the Chinese astronomers from 26 November and by the Koreans from 30 November. In Central Europe, Johannes Kepler and Johann Baptist Cysat were the most influential observers.

The length of the tail rapidly increased significantly, and soon reached questionably great, record-breaking lengths. It was reported to be between 30° and 50° on 29 November and by 10 December was between 55° and 100°! Over this period the tail was straight, but its width measured no more than about 3°.

On 8 December, Cysat, in Ingolstadt, observed the head of the comet with a telescope and noticed that the nucleus consisted of three or four parts. Between 17 and 20 December, he was able to see additional portions of the nucleus. On 24 December, however, only one bright nucleus was present, surrounded by innumerable smaller additional fragments. There is no independent confirmation of Cysat's observations, which were the first to be carried out with a telescope. Image defects could have led to this impression.

However, it is also equally likely that actual fragmentation of the nucleus was observed, because the size of the comet's head increased at the same time.

Before Christmas 1618 (in the Gregorian calendar) the brightness and tail-length declined. Although determined to be 30° on 25 December, the length decreased to less than 10° in January. Cysat succeeded in making the last observation with his telescope on 22 January 1619.

Background and public reaction

C/1618 W1 was, by the date of its discovery, the last of the three bright comets that were observed in 1618, but it was also the most conspicuous. The first comet of the year, C/1618 Q1 (1618 I), appeared in the sky in August and September. Its perihelion occurred on 17 August; the closest approach to Earth at 0.52 astronomical units was on 20 August. By the time that comet had vanished, C/1618 V1 (1618 III) appeared on the scene. It was visible in the morning sky in November and December – at the same time as C/1618 W1. Its perihelion passage had already taken place on 27 October; its closest approach to Earth at 0.17 astronomical units was on 18 December.

The fact that in November and December 1618 two bright comets were seen in the sky at the same time unleashed great disquiet in the European general population. With over 300 broadsheets published, the comets of 1618 were responsible for the peak of comet hysteria in early modern times. Their appearance was interpreted as a harbinger of the long-drawn-out Thirty Years' War, which, after years of religious and political tensions, began with the Defenestration of Prague in May 1618.

While the Aristotelian view that comets were exhalations of the Earth that were ignited by the Sun, was still in general circulation, a few astronomers doubted this view. In particular, Johannes Kepler gained the impression, from his own observations, that they must be objects in the 'supralunar' region. He had made the first telescopic

observation of a comet on 6 September 1618 with C/1618 Q1. Kepler believed that both of the two comets occurring in November had a common origin, as they both appeared in nearly the same place.

The Italian astronomer, Orazio Grassi, also saw comets as lying beyond the Moon. He believed that they lay between the latter and the Sun, because the speed of their motion across the sky was between those of the two celestial bodies, and no angular displacement through a parallax effect was observed. Galileo Galilei, however, countered this view in a polemic; he was sure that comets were of terrestrial origin and that they left the Earth on straight-line paths. This reflected unbalanced observations of the comets after perihelion, when, because of their nearly straight paths and the slow decline in brightness, they could give rise to the impression that the Earth was the origin of their motions.

Galileo also believed that a comet's tail was simply reflected sunlight, and that curved tails only arose because of refraction in the Earth's atmosphere. Cysat also held the view that comets were not self-luminous, but simply reflected light from the Sun, because in his opinion they resembled sunspots.

An allegorical depiction of the comet as a symbol of the war in Europe. While the astronomer in the foreground continues to observe, the Grim Reaper wreaks havoc.

The Great Comet of 1618 over Augsburg. *Anonymous*

Schaden gewarnet/ und bey Zeiten der Gnaden von Sünden abzustehen/ und zu der Göttlichen Barmhertzigkeit zu fliehen angemahnet würde.

Unnd ist einmahl dieser Comet ein rechter Vorbott gewesen der künfftigen Straffen Gottes/mit welchen wir heimgesucht und gezüchtiget werden sollen.

Es haben die Alten von den Cometen gesagt: Daß nie keiner erschienen/der nicht groß Unglück mit sich gebracht habe. Und Claudianus sagt von ihrer Würckung also:

——— Bella canunt , ignes subitosque tumultus,

Et clandestinis surgentia fraudibus arma,

Civiles etiam motus cognataque bella

Significant.

Und Pontanus :

Ventorum quoque certa dabunt tibi signa Cometæ ,

Illi etiam belli motus , feraque arma minantur,

Magnorum & clades populorum & funera Regum.

Das ist :

Krieg/ Auffruhr/ Blutvergiessen viel/

Dir ein Comet verkünden wil :

Unter den Leuten grosse Noht/

Auch grosser Herrn und König Todt.

Andere mehrere alte Astrologi schreiben/ daß er bedeute erstlich violenta & superba consilia, dissidia, proditiones & rebelliones, grausame und übermühtige Rahtschläge / Uneinigkeit/ Verrähteren und Auffruhr : Darnach latrocinia & subsessiones viarum , sollicitudinemque & anxietatem animorum: Das ist: Rauberey/ Unsicherheit der Strassen/ und grosse Angst und Schwermütigkeit unter den Leuthen. Zum dritten: Regum & Principum interitum, bella, pestem & morbos varios. Das ist: Grosser Königen/ Fürsten und Herrn Untergang/ Krieg/ Pestilentz und mancherley Kranckheiten. Endlich und zum vierdten/ Religionis, Legum & Institutorum mutationem, novarumque rerum inexplebilem cupiditatem. Das ist: Veränderung der Religion/ Gesetz und Weltlicher Ordnung/ beneben einer unersättlichen Begierde zu allerhand Newerungen. Wann auch die Straalen ex domo carceris herfür gehen/deutet es auff eine violentam eruptionem oder gewaltige Außbrechung und Fortpflantzung einer Lehr/ so zuvor gleichsamb als im Gefängnuß gehalten und gedrucket gewesen. Welcher Gestalt nun dieses alles in den folgenden Zeiten sich verificiren werde/ müssen wir GOTT und dem eventui heimbgestellet seyn lassen/ und wollen wir nur diß hiervon noch andeuten/ daß etliche Astrologi/ es habe sich dieser Comet in domo Religionis verlohren/ consequenter seine Endtschafft darinnen erreichet / observiret / dessen Bedeutung uns der Außgang weisen wird : Und lassen wir es hiemit bey obangedeuter Gnade/die er nach außgeschlagener Ruhtenmit sich bringen werde/verbleiben.

Es haben sich bald nach Anfang dieses 1618.

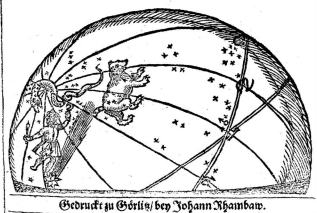

METHODUS COMETÆ
PRACTICA.

Dieſer Comet oder Wunderſtern/
iſt am vnd in firmamento erſchienen/ im
anfang des Novemb. vnd iſt endlich den
16. Ianuarii, An. 1619. widerumb
verſchwunden.

Geſtellet

Durch FRANCISCUM WENDLERUM,
Philoſophiæ ac Medicinæ Doctorem.

Pſal. 45.

Venite & videte opera DOMINI, quæ poſuit
prodigia ſuper terram.

ΕΙΚΩΝ ΙΕΡΟΓΛΥΦΙΚΗ.

Gedruckt zu Görlitz/ bey Johann Rhambaw.

▲ A chart of the comet (*above*) and title page of a treatise (*right*).
C. Hersbach (above), F. Wendler (right)

◄ The Great Comet of 1618 over Heidelberg. Detail from *Theatrum Europaeum* by *M. Merian*.

▼ Augsburg is again the scene for the Great Comet of 1618. *L. Schultes*

☄ Great Comet of 1664

When yet another bright comet appeared after the privations and horrors of the Thirty Years' War, the calls for repentance and prayer in the numerous tracts and broadsheets fell on fertile ground. This time science had little to oppose this view, given that ideas about the nature and path of the comet were so varied.

View of the Great Comet of 1664 by an unknown Dutch artist.

Data	
Number:	7
Designation:	C/1664 W1
Old designation:	1664 I
Discovery date:	17 Nov 1664
Discoverer:	Unknown
Perihelion date:	4 Dec 1664
Perihelion distance:	1.0255 AU
Closest Earth approach:	29 Dec 1665
Minimum Earth distance:	0.1699 AU
Maximum magnitude:	−1
Maximum tail length:	40°
Longitude of perihelion:	310.7°
Longitude of ascending node:	86.1°
Orbital inclination:	158.7°
Eccentricity:	1.0

All dates are in the Gregorian calendar.

Orbit and visibility

The Great Comet of 1644 was first discovered 18 days before perihelion. Because, however, the latter took place on 5 December outside the Earth's orbit, almost opposite the Sun, the result was that the comet followed an unusual path through the winter constellations. Closest approach to Earth took place just three weeks after perihelion at a distance of only 25 million kilometres.

At discovery, the comet had already passed its smallest separation from the Sun, at 11° on 7 October. Towards the end of November it

was in Corvus in the morning sky. Its path took it out of the twilight at the beginning of December into the dark skies during the second half of the night. At the same time the declination decreased, so that, for Europe, the comet only reached low heights above the horizon. The tail, however, was directed towards the northwest and was thus well-placed for observers in the northern hemisphere.

At its closest approach to the Earth on 29 December, the comet was south of Sirius in Canes Major, where it was visible throughout the night. On 2 January 1665, the Full Moon interfered with observation of the tail, which was now directed towards the north. At the beginning of January, the comet moved out of Lepus, past Rigel, and by the middle of the month crossed the eastern area of the constellation of Cetus that was traditionally shown as the monster's mouth. In doing so, it had moved into the evening sky. The last observations in March were in the constellation of Aries, when the comet rapidly moved towards its second encounter with the Sun, which it reached on 11 April, with a separation of 15°.

Many sources give inaccurate positions for the comet, which are too far to the north.

Discovery and observations

According to an analysis made by Alexandre Guy Pingré in the eighteenth century, the discovery of the Great Comet of 1664 should be credited to Spanish observers. Who precisely reported the first sighting can no longer be established, only the date, 17 November, being known.

Among the observers of the comet were Giovanni Domenico Cassini in France, Johannes Hevelius in Poland, Isaac Newton and Robert Hooke in England, as well as Christiaan Huygens in Holland. Most observations began in the middle of December, when the comet was

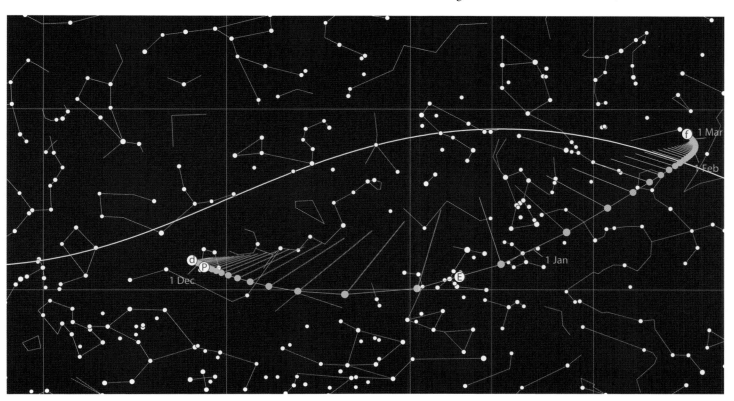

in Corvus, and was already at a magnitude of 2. A tail with a length of 10° to 12° was reported at this time.

At the turn of the year, the magnitude increased significantly. On 22 December, an observer in Italy noted that the comet shone 'brighter than every star, other than Sirius'. According to modern analysis, the magnitude must have amounted to about 0 to –1. Because the comet was outside of twilight, its brilliance was seen to best advantage in the dark night sky.

The tail was frequently marvelled at. French observers described it on 25 December as reaching halfway as far as Procyon, which would imply a length of 25°. On 27 December, Newton estimated the length as 35°. In the reports, widely differing values were often given for the length, because only the first half of the tail appeared very bright. In addition, the low height above the horizon affected the view for observers located in the north. Particularly notable is a description that states that its colour was blue.

After the high point at the end of December, the Moon interfered with detailed views. In January the magnitude and length of tail decreased significantly. On 10 January 1665, the latter was recorded as just 15°, and by the beginning of February this value had decreased to 5°. Until 18 February, the comet was visible to the naked eye. The last observations were telescopic ones on 20 March.

Background and public reaction

The Great Comet of 1664 unleashed a flood of well over 100 texts and tracts in German-speaking countries. Most of these broadsheets called upon readers to repent and reform. The fuss was only amplified by another comet at Easter 1665. C/1665 F1 passed much closer to the Sun, at a distance of just 0.11 astronomical units, than C/1664 W1. It was followed between 27 March and 20 April, and had a magnitude of up to –1, only shortly before the plague broke out in London in May 1665. By the end of the year, 90 000 people died there alone.

The relatively straight path of C/1664 W1 across the sky led many to believe that the comet was moving on a linear orbit. Hevelius was of this opinion, but at the same time envisaged the comet's orbit as a conic section, with the Sun at the focus. Cassini, however, thought that the comet was orbiting Sirius, while the star and the comet moved together around the Earth – Cassini was a resolute opponent of the Copernican world-view. Against this, Giovanni Alfonso Borelli recognized correctly that the comet of 1664 was following a curved, heliocentric orbit, similar to a parabola. Adrien Auzout, who also believed in an elliptical or parabolic orbit, even calculated, on the basis of his own observations between 22 and 31 December, an ephemeris for its future path against the sky – the very first of its kind.

The possibility of calculating the orbits of comets more accurately and of comparing them with other apparitions, was, however, not taken for granted in 1664. The French scholar Pierre Petit was of the opinion that the comet of 1664 was the same as the one that appeared in 1618, and predicted a return in 1710. Because of the failure of his prediction, he was later known as 'Stolperstein' ('stumbling-block'),

a word-play ('Stein') on his first name, Pierre, which also means 'stone' in French.

Neither was there any frequent agreement over the physical nature of the comet. Hevelius believed that comets were disks that always turned the same side towards the Sun, and arose in the atmospheres of Jupiter and Saturn. Hooke, on the contrary, had observed that, in the telescope, comets did not show any phases like Venus, and thus concluded that they must emit their own light. Almost prophetically, he continued by saying that the tail consisted of individual cometary particles, that were driven off by rays from the Sun.

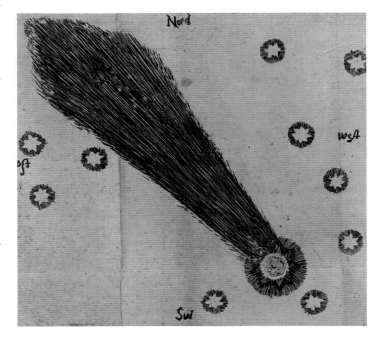

The Great Comet of 1664 (*above*: detail from a broadsheet) was claimed to be responsible for the landslide at Geißlingen on 15 Dec. 1664 (*below*). *J. T. Theyner (above), C. Mayer (below)*

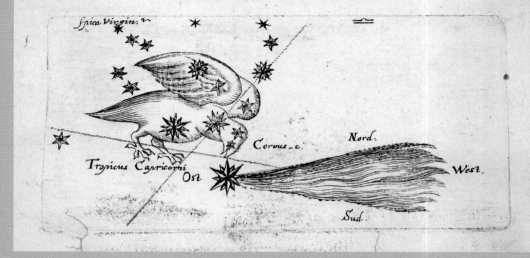

Eigentliche Abbildung / des grossen Cometen / so den 18.
Dec. St. Nov. 1664. zu Augspurg / von vielen glaubwürdigen Personen
gesehen / und dann ferner zu Nürnberg / Straßburg / Hamburg / Lübeck /
Leipzig und dergleichen mehrer Orten sich sehen lässet /
dessen Bedeutung ist Gott bekant.

◄ A diagram of the location of the comet in the constellation of Corvus (left). *Anonymous*

▼ Daily changes in the appearance of the comet according to *J. Hevelius*

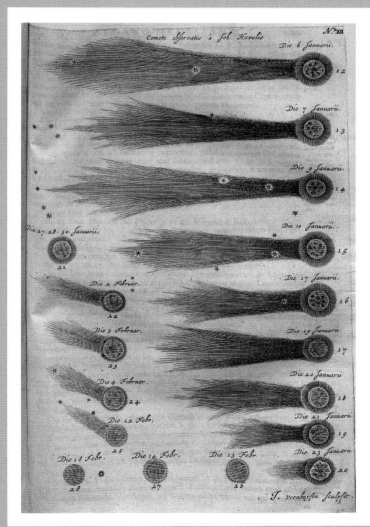

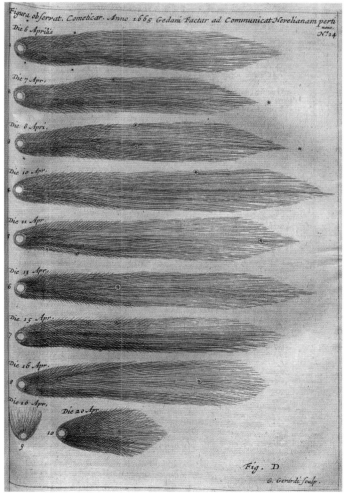

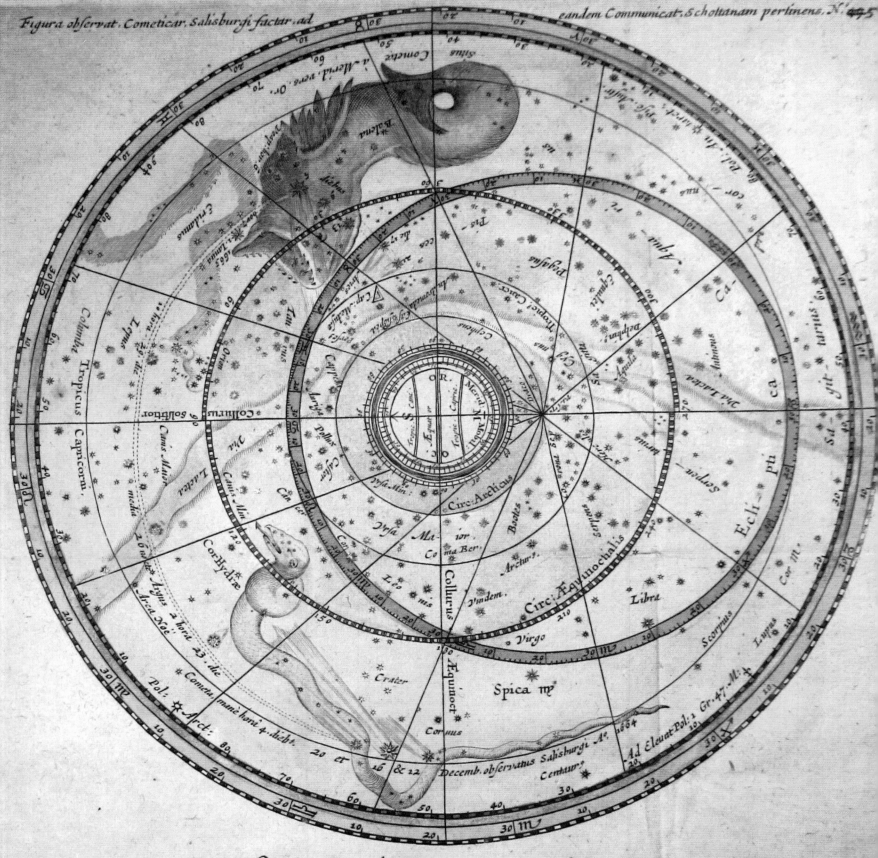

SYSTEMA ASTRONOMICVM COMETÆ

Mense Decembr. A°. 1664. et Ianuar. A°.1665. Salisburgi observati.

Locus Cometæ. 12 Decemb. in Cauda Hydræ. 16. Eiusd. supra inferiorem partem ventris Hydræ. 20. Eiusd. in ventre Hydræ. 23. eiusd. subtus Cor Hydræ. 26. Eiusd. in Via Lactea. 29 Eiusd. in Capite Leporis. 1. Ianuar. 1665. prope Eridanum. 5. Ian. in pectore Balenæ. 9. Ian. in Collo Balenæ. 13 Ian. in Occipite Balenæ. 17. Ian. circa pectus Arietis. 21. Ian. circa Pisces, versus Andromedam.

Altitudo Cometæ ab Horizonte elevati. 12. Dec. 13. gr. 16. Dec.15. gr. 20. Dec. 18. gr. 23. Dec. 20. gr. 26. Dec. 23. gr. 29 Dec. 26. gr. 1. Ian. 27. gr. 5. Ian. 29. gr. 9. Ian. 37. gr. 13. Ian. 45. gr. supra Æquat. 4. gr. 17. Ian. 55. gr. supra Æquat. 13. gr. 21. Ian. 59½. gr. supra. Æquat 17½

Declinatio Cometæ à Meridie vers. Orientem. 12. Decemb. 34. gr. 16. Dec. 24. gr. 20. Dec. 14. gr. 23 Dec. 11. gr. 26. Dec. 9. gr. 29. Dec. 6. gr. 1. Ian. 23. gr. 5. Ian. 19½ gr. 9. Ian. 10. gr. 13. Ian. 5. gr. 17. Ian. 2½ gr. 21. Ian. sub Meridiano.

Cauda Cometæ in Hydra existentis, versus Cor hui9 syderis in Occidentem protensa, Extra Hydram à Luna obfuscata. In Balena, intra utru9 caput Medusæ et Tauri Septentrionem versus prominens.

Imago Cometæ, in Hydra existentis, supra Meridianum, Coloris Iovialis; prope et infra Meridianum Saturnini. Extra Hydram Coloris Saturnini. Balena, sub eius Ortu, Coloris Martialis; prope et infra Meridianum Mercurialis; remotior tandem, Saturnini.

Motus Cometæ Sphæricus et Spiralis, sub singulis horis matutinis 15. gr. 45. minutas, sub Vespertin. 23. gr. in hoc Climate superans, sed retardat, r Asterismus hui9 Cometæ ex angulari Planetarum distantia elicitus; intercidit Guttur Vulturis volantis et Cygni; Dorsum Draconis; et Sinciput Maioris Vrsæ

Abbildung
deß Neuen
Comet=und Wunder=Sterns/

Wie sich derselbe in den Innern Oesterreichischen Landen/ und benachbarten Croatischen Orten/ besonders aber über Rackelspurg und Czackenthurn Morgends zwischen 2. und 3. Uhren den 12. Januarii dieses 1664sten Jahrs/ mit erschrecklicher Entsetzung der Anschauenden/ hat sehen lassen.

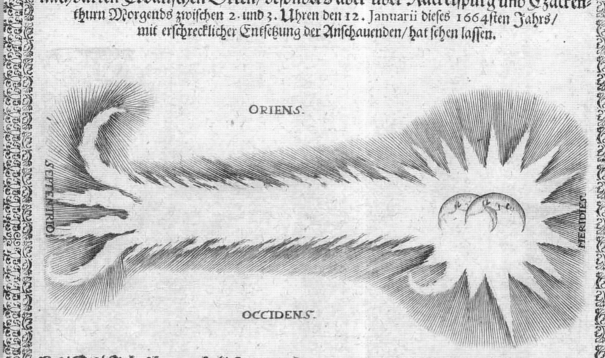

ORIENS.

SEPTENTRIO

MERIDIES

OCCIDENS.

Jese übernatürliche/ und fast niemaln dergleichen am Himmel/ in so entsetzlicher Gestalt/ und brennenden Feuer/ sich sehenlassende/ angezündete Göttliche Zorn=Fackel/ ist vermuthend viel grösserer Kriegs=Wurckung/ und anderen Unheils/ als dessen so Anno 1618. sich 30. Tage über gantz Europa gewiesen/ und darauf der blutige dreyssigjährige Krieg/ wie auch desselbigen Anno 1652. auf welchen der Polnische/ Schwedische und Dähnische Krieg erfolget/ hoch zu besorgen. Dahero/ weil die Form solches Wunder=Sternes so über groß verwunderlich/ besonders die darin erscheinende halbe Monden zu ersehen/ dörffte/ weil bey itzigem Zustand/ die von dem Erb=Feind dem Türcken übermächtige Rüstungen durch gantz Europa erschollen/ gewiß nichts gutes predigen/ weßwegen dem erzörneten GOtt durch Buß= und eyfrigem Gebet allzu grosse Zeit in dessen Ruthe zu fallen ist/ der Höchste wende alles Ubel. Und weil sonderlichen in deß Neubarthi ausgefertigten Calendario deß verstrichenen 1663sten Jahres unterm Titul die grosse Conjunction, der beyden Obern Planeten Saturni und Jovis im feurigen Zeichen deß Himmlischen Schützen aus solcher erfolgende Comet= oder Wunder=Stern gemuhtmasset/ ist wohl Beyfall zu geben/ daß solcher von selbiger Conjunction entsprossen. Also dörfften wohl einige Astronomische Gutachten/ wegen Wurckung solchen Wunderbothens/ ans Liecht gebracht/ und alsdann durch den Druck zu männigliches Nachricht mitgetheilet werden. Soviel in Eil zur Nachricht.

Zufinden in Nürnberg bey Paulus Fürst/ Kunsthändlern.

Chart of the path of the comet as observed at Salzburg (*opposite*). Broadsheet about the appearance of the comet above Radkersburg (*above*). G. Gerrits (*opposite*), P. Fürst (*above*)

⚡ Comet Kirch 1680

Plague in Europe, and the Turks at Vienna: the greatest comet of the seventeenth century appeared at a turbulent time. Discovered and followed with the telescope, many experts took its two apparitions before and after passage past the Sun as two different comets. An amateur from Saxony and an English scientific genius recognized that parabolic orbits could unite these apparitions. The man on the street, however, knew nothing about this, and was frightened by the 'terrible' comet.

Data	
Number:	**8**
Designation:	**C/1680 V1 Kirch**
Old designation:	**1680**
Discovery date:	**14 Nov 1680**
Discoverer:	**Johann Gottfried Kirch**
Perihelion date:	**18 Dec 1680**
Perihelion distance:	**0.0062 AU**
Closest Earth approach:	**30 Jan 1680**
Minimum Earth distance:	**0.4895 AU**
Maximum magnitude:	**−10**
Maximum tail length:	**90°**
Longitude of perihelion:	**350.6°**
Longitude of ascending node:	**276.6°**
Orbital inclination:	**60.7°**
Eccentricity:	**0.99998600**

All dates are in the Gregorian calendar.

Orbit and visibility

When Kirch discovered the comet near the Moon and Mars, it was close to the ecliptic in Leo in the morning sky. Initially, it moved, like the Moon and planets, east along the ecliptic towards Virgo. At the end of November 1680 it passed the constellation's principal star, Spica. The period of visibility in the morning sky shortened in the following weeks, in which C/1680 V1 approached the Sun. By 9 December, the comet had reached 20° from the Sun and was invisible.

Nothing was seen of perihelion on 18 December, when the comet reached the extraordinarily close distance of less than 500 000 km from the surface of the Sun. It was not until 20 December that the comet appeared in the evening twilight. It rapidly left twilight and was soon seen in dark skies.

Its best visibility now followed, although there was interference from the Moon. By Christmas 1680 (according to the Gregorian calendar), the comet was in the constellation of Aquila. Its tail stretched towards the north in the Milky Way, rising at right-angles to the horizon. In January 1681, conditions improved even further. The comet was then coming nearer and nearer to the Earth, and at the same time, it rose higher above the horizon. At the beginning of the month, it was in the constellation of Pegasus, at the end of January in the region of Pisces, Andromeda and Triangulum. It was there that closest approach to Earth occurred on 30 January. The comet eventually disappeared towards the east in the constellation of Perseus, where it was seen for the last time in March 1681.

C/1680 V1 belongs to the sungrazers, but is not a member of the Kreutz Group, which has produced many bright comets.

◀ The Great Comet on 22 Dec. 1680. Watercolour. *Rochus van Veen*

Discovery and observations

Johann Gottfried Kirch discovered the comet on the morning of 14 November 1680, when in Coburg, after he had observed the Moon and Mars through his telescope. Kirch thought that it must be 'either a comet or a nebulous star, similar to the one in Andromeda's belt'. The comparison with M31 suggests that the comet did not appear very faint, and must have been at least magnitude 4.5. Indeed, Kirch could see the comet with the naked eye, but it was, however, almost hidden by the light from the nearby Moon.

By 16 November a small tail had already become visible through the simple telescopes available at the time. By 21 November this had grown to a length of half a degree. The brightness amounted to about magnitude 2 at this date. The comet then brightened rapidly. By the end of November it had blossomed into an impressive sight with a tail between 15° and 30° long. The tail was, however, extremely faint. Kirch himself stated that he had never seen a comet with such a faint tail. The brightness of the head, however, increased to magnitude 1 by December, and must have increased still further – although because of its closeness to the Sun, the comet was not visible.

After perihelion on 18 December individual sightings in daylight were even recorded. The comet was, for example, seen from the Philippines on 18 December, only 2° away from the Sun, and was seen on the following day by settlers in New York 'slightly above the Sun'.

On 20 December, the comet was rediscovered in Europe in the bright evening twilight. The head, still too close to the Sun in the bright twilight, was not visible. However, it projected a narrow, long tail right up to the zenith, and, as twilight progressed, it emerged ever more clearly.

In subsequent days an incomparable spectacle unfolded: an extremely bright tail stretched north from a red, glowing cometary head, low on the horizon. It reached from Scutum across Aquila and Cygnus as far as Cassiopeia, "so long, that it stretched right from one side of the horizon to the other, but with little change in its width", as Casimiro Diaz reported from the Philippines. The tail was still visible four hours after the head had set, and according to Diaz, was "giving as much light in the darkness of the night as a Quarter Moon". Much admired was the golden colour of the tail, which, at this stage, appeared straight, the width of which did not exceed 3°, but near the head measured just 0.5° across.

John Flamsteed described the nucleus as reddish, seen through a telescope, probably because of its proximity to the horizon. To him it appeared square in shape. His compatriot Robert Hooke reported a tail length of 90° and a width of 2° on 28 December. On that occasion he noticed a ray of light that seemed to escape from the head of the comet "like the sudden spurt of water from a machine" – perhaps the first hint of a jet from a cometary nucleus.

By the turn of the year, 1680–1681, the comet had become fainter. On 31 December the head must already have reached magnitude 3. The tail, however, remained impressively long into January at 55° to 75°. It was now slightly curved towards the south. On 7 January, Kirch was able to observe the phenomenon of an anti-tail, which he described as an 'anal tail'.

Flamsteed described the subsequent development of the comet. According to him, on 13 January, 25° of the tail was still to be seen, the brightness of the head corresponded with that of a 4th-magnitude star. On 19 January, it was still magnitude 5, and 7 on 7 February. While the head had already been lost to the naked eye, the tail remained visible until the middle of February. Georg Samuel Dörffel was the last to catch sight of it on the 17th of the month.

The last determinations of position in March 1681 came from Isaac Newton, who hoped to obtain as many positions of the comet as possible with his telescope. He succeeded in being the last person to see the Great Comet of 1680 on 19 March 1681.

Background and public reaction

Kirch's Comet was the very first comet to be discovered with a telescope. After its peak phase in December 1680, it was followed more intensely than any previous comet, having been observed with telescopes. The new instruments demonstrated their superiority over observations with the naked eye, because it was now possible to follow the comet far longer and thus obtain more evidence about its path across the sky.

Frequently, the two appearances of this comet before and after solar conjunction were taken to be two different objects. Newton and Kirch, the most prominent observers, were themselves convinced by this idea. The theologian and amateur astronomer from Saxony, Georg Samuel Dörffel, however, demonstrated in his work *Astronomische Beobach-*

tung des Großen Cometen ('Astronomical observation of the Great Comet') that both phenomena were related to a single object. He proposed a parabolic orbit as a solution – he was led to this form of orbit by simple estimates of the angle between the comet and the Sun at perihelion and the acceleration at perihelion.

John Flamsteed, too, was convinced that it was not a matter of two comets on straight-line orbits, as Newton maintained. He assumed that magnetic forces from the Sun were responsible for the 'curved' orbits. He mistakenly ascribed perihelion to have taken place on the near side of the Sun, rather than on the far side. Flamsteed believed that comets consisted of water, which turned into vapour under the rays of sunlight – an astoundingly modern concept!

For a long time Newton doubted the reality of a non-linear orbit, even though he had recognized that comets must be illuminated by the Sun, because they appeared brighter around perihelion than when near the Earth, and were obviously brighter after perihelion than before. He finally decided, independently of Dörffel, on a parabolic orbit, which he published in 1687 in his famous *Principia*. Cometary orbits could therefore be calculated and predicted, if enough observations were available. In the same year, Edmond Halley realised that predictable orbits could also serve to identify former cometary apparitions with current ones. He later considered the comet of 1680 to be a return of the comets of 44 BC, 531 and 1106, and predicted that it would return in the year 2255. Leonhard Euler, on the other hand, in the eighteenth century, calculated a period of 117 years, whereas Alexandre Guy Pingré took a figure of 16 years, and Johann Franz Encke even one of just 8.8 years.

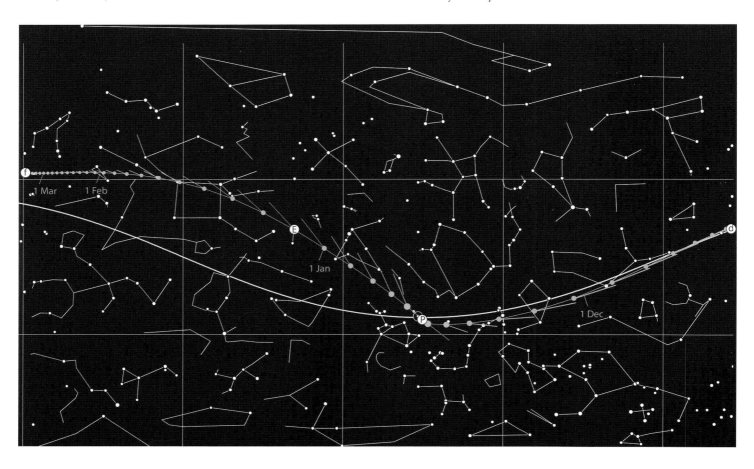

Path of the comet from the former constellation of Antinous (the modern-day Scutum) to Pegasus. The tail length depicted amounts to about 60°. The angular depiction of the comet's path is notable.
Anonymous

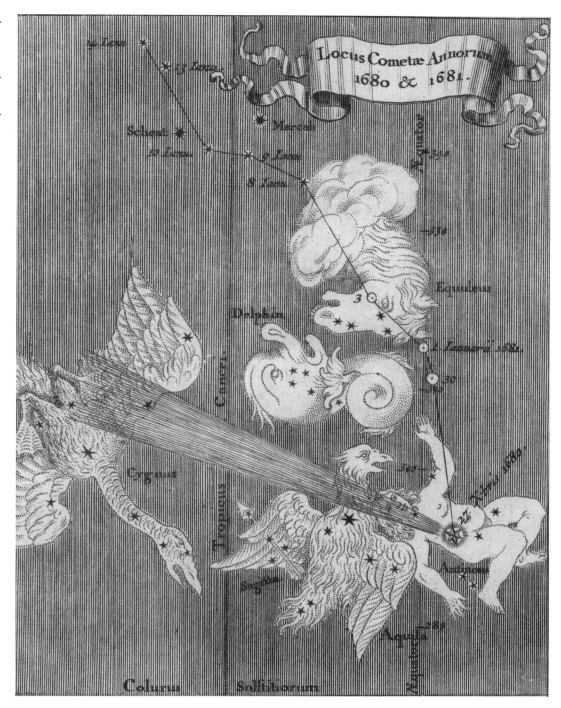

Despite the advances in scientific research that accompanied the comet of 1680, comet hysteria in early modern times reached its last significant peak. More than 230 tracts and broadsheets appeared in Germany. In most of them the comet was interpreted as a warning sign from God. This definitely had an effect on everyday life. The council of the free city of Regensburg, for example, decreed that 'as God, in his outrage, and with justifiable zeal has set a warning in the firmament to bring it home to us, as a scourge of dreadful retribution and wrath, all sinful behaviour must be renounced with remorseful and penitent behaviour.' The citizens were instructed 'to relinquish all pernicious ostentatious clothing and all overweening pride, to permit no private dancing, to abstain from mummery and inappropriate sleigh rides, both in small and large sleighs, to completely abstain by day and by night, and to make a great effort towards a calm and sober change of behaviour.'

Increasingly, however, the demystification of comets was taking place. Erhard Weigel, one of the teachers of the discoverer, Kirch, himself wrote eight essays against the fear of comets. Jacob Bernoulli also loudly denounced the misinterpretation of these astronomical objects.

In many broadsheets, however, the predictions calling for repentance took precedence over entertainment. The stereotypical 'fearfulness' produced a calculated spine-chilling account, because the broadsheets were a lucrative business for the authors and publishers, and frequently the way in which they earned their living. The desire to scare people was, however, a sign of the uncertain times: an epidemic of the plague raged between 1679 and 1684, and the Turks were once more at the gates of Vienna.

Vernünfftige
Erkantniß und eigentliche Bewandniß/
Deß den ¹⁶/₂₆. Decembris erstesmal zu Abends um 1. Uhr der Grössern allhier zu Nürnberg erschienenen entsetzlichen
Cometens.

Gedruckt/ im Jahr Christi/ 1680.

The Great Comet of 1680 over Nuremberg, depicted from similar points of view when above the western horizon. *Anonymous (above), J. J. Schollenberger (opposite)*

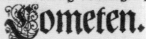

Abbildung und Beschreibung
deß wunderwürdigen unvergleichlichen
Cometen.
Der erstmals zu Anfang deß Wintermonats vor Aufgang der Sonnen erschienen / und anjetzt nach
derselben Untergang sich entsetzlich sehen lässet.

MAn findet sowoln in Heiliger Schrifft / als auch andern glaubsichern Historien / daß so offt der Allmächtige GOtt die Sünden einiger Erd-Einwohner / zu bestraffen / sich entschlossen / Er solches zuvor / aus mild-vätterlicher Langmut entweder durch wahre Propheten / oder entsetzliche Wunder ankünden lassen : Hat sothane Warnung gefruchtet / und ist eine eyferige Buß und Bekehrung erfolget / so ist auch die angedrohete Straffe abgewendet / im widrigen aber / unfehlbar vollzogen worden.

Wann nun von einigen Jahren her der erzörnete und darbey gütige GOtt / durch Erdbeben / unterschiedliche Cometen / und andere mehrfältige ungewöhnliche Zeichen / denen in Sünd und Lastern blindersoffenen Menschen / ihre wolverdiente annahende Jammer-Plagen zur Genüge anmelden / auch hinterbliebene Besserung aber / durch verderbliche Krige / Sterben / und anders Elend / (wie man leyder mehr als zu viel erfahren) eyfergrimmig ergehen lassen.

Als hat Er nun abermaln an dem hohen Himmel / eine erschröckliche Fakel / Ruthe und Schwerdt / zu einer gütigen Warnung / für den annoch bevorstehenden Unglük aufgesetzet ; Damit weiln je alles vorige im Wind geschlagen worden / dieser grausame förchterliche / wegen seiner Gestalt und Lauffes / von denen Gestirn-Erfahrnen unvergleichlich bewunderte Comet / einige Entsetzung / und Veränderung in den Sünd-verstockten Gemüthern auswürcken / und die nunmehr abgeurtheilte Sünden-Straffen / durch herzliche Reue und Buß / zurük gezogen werden möchten.

Es ist aber dieser wunderwürdige unvergleichliche Comet / allhier und auch anderer Orten / das erste mal / in dem Zeichen deß Löwen / worinnen sich auch damaln der Kriegs-Planet Mars befunden / unter deß Löwen Herz-Stern / an dem Zodiaco oder Thier-Kreise und Planeten-Wege Südwerts streichende / sodann folgenden Tagen in dem Zeichen der Jungfrauen / Anfangs / ohne / hernacher den 12. Novembris mit einem immer zunehmenden / doch wegen anbrechenden Tages schwachliechten Schweiffe gesehen worden ; ist hierauf wegen seiner Annäherung zur Sonnen / als auch eingefallenen trüben Wetters / unserm Gesicht entwichen / und so lang unsichtbar geblieben / biß er endlich / nachdeme er die Sonne überlauffen / und sich aus deren Glantz erlediget / auch die Luft sich wieder ausgehellet den 16. (26) dieses innstehenden Christmonats / bey angehender Nacht / mit einem sehr langen blaßweisen Schweiff ganz prächtig hervor gebrochen / und sich denen Erd-Einwohnern / als ein Rach-Schwerdt und Zorn-Ruthe deß Allerhöchsten GOttes / entsetzlich vor Augen gestellet / hat sich damaln befunden / in dem ungebildeten Zeichen deß Steinbocks / zwischen der Sonnen-Strassen und Welt-Gürtel / unter deß Adlers Knaben auf dem Altar : Zeither / hat er seinen vorher denen andern Planeten gleich-geführten Lauff / mit grosser Verwunderung der Stern-Erfahrnen dergestalten geändert / daß er nicht allein ruckgängig worden / sondern auch sich immer höher gegen Norden erhebt / auch bereits die Tag und Nacht gleichende Linie überstiegen / und da er wie bisher geschehen / also fortfahren solte / uns bald vertical werden / oder über das Haupt kommen dürfte. Er hat nunmehr von seiner ersten Erscheinung an / den Löwen / Jungfrau / Waag / Scorpion / Schützen / auch mehrentheils deß Steinbocks / und also sechs Zeichen / ja dergestalten den halben Himmels-Bezirk durchgelauffen / und wird seine jetzig vier und zwanzig stündige Bewegung über drey Grad vermerkt. Von der wahren Grösse / sowoln seines Cörpers als Schweiffes / ist darum / weiln seine Höhe nicht eigentlich bewust / nichts unfehlbar Gewises zu gedenken. Es ist aber gleichwoln aus unterschiedlichen denen Sternkündigern bekannten Ursachen zu vermuthen / daß er höher als die Sonne stehen / auch daher / ob er schon unsern Augen kaum als ein Stern der dritten Grösse vorkommet / derselben in der Grösse nichts bevor geben möchte / und wird also wann man den Bezirk seines Cörpers / auf mehr als tausend / die Länge deß Schweiffes aber / auf etlich mal hundert-tausend Teutsche Meilen schätzet / nichts der Warheit Ungemässes begangen. Hingegen aber irren die jenige / so da fürgeben / daß der Schweiff welcher auf sechzig Grad lang auzusehen / und deren Eins auf unserer Erden gerechnet fünfzehn Meilen gibt / auf tausend Meilen lang seyn müsse ; dann weiln augenscheinlich bewust / daß unser Wohnhaus diese Erdkugel / gegen den unmäßlich weiten Umkreysen der Planeten / und anderer höherer Gestirne kaum vor ein kleines Pünctlein zu achten / und die auf Erden 15. Meilen lange Graden / sich auf viel tausend ja hunderttausend vermehrfältigen / so muß auch der Comet als ein in dergleichen Kreyß stehendes Gestirn / nicht aber als ein auf Erden kriechender Wurm betrachtet / und seine Grösse berechnet werden.

Der allergrösseste HERR / der uns dieses grosse Warnungs-Zeichen von den hohen Himmel leuchten lässet / wolle unser aller Herzen und Gemüter dergestalten regieren und führen / daß wir in Herzens-Reue / wahrer und beständiger Bußbekehrung / Ihme mit eyferbrünstigen Gebet demütig zu Fussen fallen / und die durch diesen Cometen angedrohete schwere Straffen abbitten und abwenden mögen ; Wollen daher mit folgenden Sonnet schliessen.

O Ist / Lust / Laster-voller Sünder! wilst du nicht von Scham erröthen /
so erbleiche nun aus Schrekken! Sih' auf / wie dich übersteigt /
diese bleiche Feuerrakkete / blik an / wie der höchste zeigt /
deiner Sünden Straffruthen / merk wie von dem Kriegs-Planeten
kommt ein Schwerdt auf dich gelauffen ; Schau den grausamen Cometen /
der mit blassen Todesstrahlen / von dem hohen Himmel leucht /
der mit gleichlos langem Schweiffe / in die Nordgestirne reicht :
Ach es redet ohne Rede / GOtt durch diesen Straff-Propheten /
Narbe / Schwerdt / Gifft stehen fertig / dich O böser Mensch zu schlagen /
mit Krieg / Armut / Krankheit / Sterben / und mit allen Jammer-Plagen /
wann du wirst verstockt beharren / in den Lastern ohne Scheu ;
wirst du aber dich bekehren / und mit Herz- und Schmerzens-Reu /
von dem Sündentod aufleben / so kanst du Vergebung hoffen /
dann die Thür der Gnaden GOttes / stehet dir noch immer offen.

Des Neuen Wunder großen Comet Sterns von West westsüd gegen Nordostseigner Lauff
Sambt der gegend und beschaffenheit des Observatory in der Basten auf der
Pösten in Nürnberg observirt und vor augen gestelt.

Donerstags den 16 Decembr (da er am grösten und auch vermuthlich in perigœo gewesen) locus ejg in 14 ♌ longitudinis 13 latit. borealis. Die 17 Decemb. locq in 17 ♌ long 15 lat. bor Die 18 locq in 20 long. 18 lat. bor die 19 hujus locq in Ecliptica 24 ♌ long 19 lat. bor. Die 20 locq in 27° 30 long. 2½ lat bor Die 21 Dec. locq in 2.30 ♏ long. 23 lat. bor. Die 22 Dec 1680 hujus in 9 ♏ long. 19 lat. bor. Plures observationes cœli Tempestas denegavit. 1 Januarioi68i

The comet on 26 Dec. 1680 over the castle of Nuremberg with the instruments of the Eimmart Observatory. Etching (*above*). Depiction of the same evening over the city with the view towards the west (*below*). The position of the Moon is erroneous. *J. J. von Sandrart (above), J. Hofmann (below)*

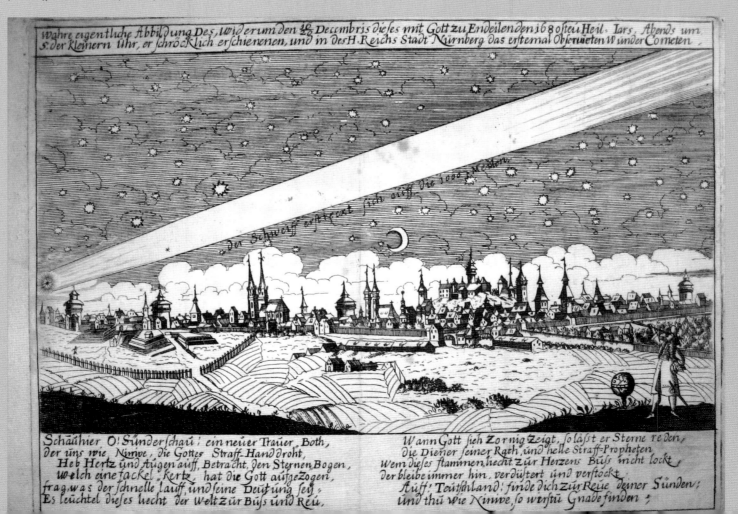

Wahre eigentliche Abbildung Des, widerum den 16 Decembris dieses mit Gott zu Endeilenden 1680 steu Heil. Iars, Abends um 8 der kleinern Uhr, er schröcklich erschienenen, und in des H. Reichs Stadt Nürnberg das erstemal Observeten Wunder Cometen

Der Schwaiff erstreckt sich auff die 100 Meilen

Schau hier O! Sünderschau, ein neuer Trauer, Both,
der uns wie Ninive, die Gottes Straff-Hand droht,
    Heb Hertz und Augen auff, Betracht, den Sternen Bogen,
    welch eine fackel-Kertz, hat die Gott auffgezogen,
fragwas der schnelle Lauff, und seine Deutung sey,
Es leuchtet dieses liecht der Welt zur Büss und Reü

Wann Gott sich Zornig Zeigt, so läst er Sterne reden,
die Diener seiner Rath, und helle Straff-Propheten
    Wem dieses flammen liecht zur Herzens Büss nicht lockt
    der bleibe immer hin, verdüstert und verstockt
Auff! Teutschland: finde dich zur Reüe deiner Sünden;
und thü wie Ninive, so wirstu Gnade finden;

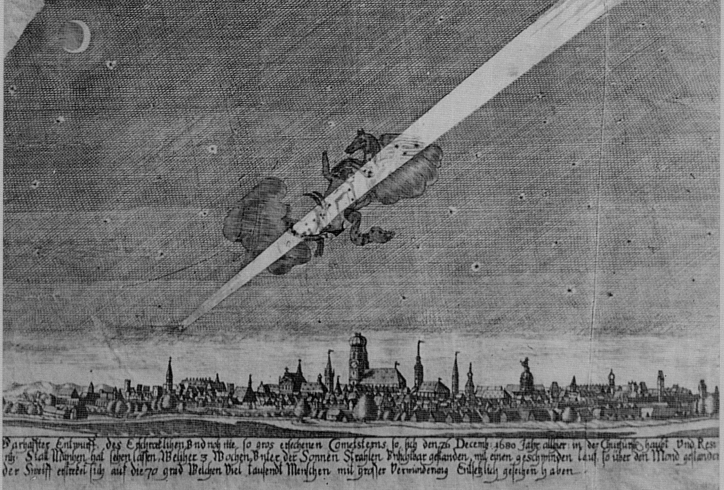

The Great Comet of 1680 over Munich (*above*) and Breslau (*below*). The constellations through which the comet passed are depicted in completely different ways. *Anonymous (above), M. Buck (below)*

Eygendlicher Abriß / des Anno 1680. entstandenen

Cometen.

ES ist derselbe zwar erstlich im Zeichen der Jungfrawen / den 11. Novemb. allhie zu Basel aber / wegen trüben Luffts / erst den 24. Novemb. Morgens gesehen worden / ein wenig vnderhalb der Mittägigen Waagschalen stehend / vnd seinen Schweiff auff die 8. Grad lang gegen dem hellen Stern der Jungfrawen Aehren genañt oder in anschung der Sonnen-lauff / gegen Nidergang derselben / vnd ein wenig gegen Mitternacht gerichtet / folgenden Tags koñte man wegen annahendem VollMoñs jhne nicht mehr sehen / biß er bey der Sonnen vorbey gegangen war / vnd den 12. Decemb. Abends widerumb herfür kam / da dann (einkoñienem Bericht nach) der Stern des Cometen zwar noch nicht / der Schweiff aber in einer grausamen Grösse erschienen / daß er vom Horizont biß zum Schätiel-puncten sich erstreckte / vnd Fewr-roth / auch so breit war / als ein Regenbogen / anzuschen wie eine hohe Saul oder Balcken / gegen Auffgang der Sonnen vnd ein klein wenig mehr gegen Mitter-nacht auffgerichtet : Allhie aber war es trüb Wetter / biß auff den 18. 19. vnd 20. Decemb. da koñte man observieren / daß der Comet von der Ecliptica oder Sonnenstraß herauff kam / vnd bey der Lincken Hand Antinoi vber die Æquinoctial-Linien / vnd auß dem Steinbock in den Wassermann gieng / seinen Schweiff streckte er in einer geraden Linien von der Sonnen hinweg vber den Delphin vnd Rechten Flügel des Schwanens gegen der Cassiopeæ auff die 70. Grad

A. Comet / wie er den 20. Decemb. Anno 1680. gesehen worden.
B. Wie den 5. Jenner / An. 1681.
C. Wie den 25. Jenner eodem.
D. Polus Eclipticæ.
E. Polus Mundi, vnd darbey Stella Polaris.
F. F. Circulus Polaris.
G. G. G. Tropicus Cancri.
H. Pars Eclipticæ.
I. Aries der gestirnte Wider.
K. Die Mucken oder Bine.
L. Medusæ Haupt.
M. Der fliegende Held Perseus.
N. Ein heller Stern Capella oder das Geißlein genandt.
O. Das Haupt Aurigæ oder Erichthonii.
P. Camelopardalis.
Q. Die stolze Königin Cassiopea.
R. Cepheus, Ihr Mann.

Magnitudo Stellar
Prima.
Secunda.
Tertia.
Quarta.
Quinta.
Sexta.
Nep.

Gestochen vnd gedruckt zu Basel des J.J. Thurneysen Kupfferstecher vnd Künsthändler auff dem S. Albangraben / Inn Ernanweysihen.

lang / er war weiß oder bläich vnd nicht mehr so bräit als zuvor / vnden bey dem Sternen etwas heller vnd dichter beysammen / Oberhalb aber etwas Dünner vnd mehr außgebräitet / jedoch endlich wider zusammen gespitzt ; Der Stern selbsten war trüb vnd klein / kaum so groß als der nicht weit darvon stehende Venus-Stern ; in den folgenden Tagen / biß auff den 5. Jenner dises 1681. Jahrs / gieng der Comet bey dem Delphin / wie auch bey der Schnurzen vnd Halß des Fliegenden Pferds Pegasi fürüber / durch dessen Brust / vnd bey dem Rechten Schenckel vber den Tropicum Cancri vnd auß den Fischen in den Wider : Ob zwar der Schweiff allgemachlich abgenommen / also daß er von dem Haupt der Andromedæ, darbey der Comet damahls allernechst stunde / kaum noch 30. Grad / biß vber den Perseum erreichte / so wurde doch der Stern nicht kleiner / sondern scheinte heller als zuvor : In seinem täglichen Lauff aber verzichtete er nur Zween ; vnd hernach nur Einen Grad / biß er den 25. Jenner mit den zweyen Sternlin in basi Trianguli gegen dem Mucken oder Bienen in dem 9. Grad des Stiers / eine gerade Linien machte / seinen kaum noch sichtbahren Schweiff nur auff 10. Grad / in das Haupt Medusæ außstreckend. Da nun ein solch schröckliches Spectaculum am Himmel / nicht auch seine sonderbahre wichtige Bedeutung / grosser Enderungen vnd vnvermuthlicher Zufählen auff Erden habe / wird kein Verständiger es leichtlich widersprechen können ; Darvon aber (Geliebts Gott) ein besonder Tractätlein erfolgen solle.

P. Megerlin / D.

S. Andromeda, Ihr angefestiete Tochter.
T. Pegasus Das Fliegende Pferd.
V. Equuleus das kleine Pferd.
W. Cygnus Der Schwan.
X. Der Drach.
Y. Die Leyr oder fallende Geyer.
Z. Der Fluß Tigris.
a. Der Pfeil.
b. Delphinus oder das Meerschwein.
c. Der Adler.
d. Der Knab Antinous oder Ganymedes.
A. B. C. Der Lauff des Cometen / da die Ringlein bedeuten / wann er hat können gesehen werden / die Pünctlein aber / wann er wegen Gewölcks oder hellem Mondscheins vnsichtbar war vom 20. Decemb. 1680. biß zum 25. Jenner An. 1681.
e. Via Lactea, Die Milchstraß.

Gedruckt zu Basel /
Durch Johann Rudolph Genath.

The passage of the comet through the constellations from 30 Dec. 1680 (A), through 15 Jan. 1681 (B) to 4 Feb. 1681 (C) (Gregorian dates). *P. Megerlin*

Jacob Milich. Comment. in Plin. lib. 2. de Hist. Mundi.

Receptum omnium seculorum consensu, Cometas prodigia esse.

Eigentlicher Bericht/ welcher Gestalten der nachdenckliche/ dises

zu End=lauffenden 1680sten Jahrs/ noch in der Lufft stehende Comet zu Marckt=Wailtingen
etliche Abend observiret worden.

Nach dem in dem Monat Decembri ein Zeitlang / sowol Tags / als Nachts / die Lufft trüb und wolckicht gewesen/hat selbige sich den 16. 26. gegen Abend anfahen aufzuhellen / daß sich das Gewülck nah und nah von Osten / gen Westen gewalket / nicht anderst als ein Vorhang auf einem Theatro weggezogen wird; Darmit was der Allerhöchste in der Lufft uns wolte weisen/auch möchte gesehen werden. Wie dann nach untergangener Sonnen/zwischen 5. und 6. Uhr / aus dem Westseits noch etwas stehendem Gewölck/unversehens ein lang=breit= und bleicher Schein / oder Strahl / sich erwisen / ein Anzeigen eines gegenwärtigen Cometen oder Schein=Sterns; so desto greßlicher und entsetzlicher anzusehen war/weilen vorher/mit das geringste (ob schon von einem Cometen aus anderen Orten bereits Bericht gethan/und deßwegen alle Nacht/ so offt die Lufft hell gewesen/ fleißige Aufsicht gehalten worden/) dergleichen können vermerckt worden. Weilen nun wie erst gedacht/die Lufft Sudwest seit / wo der Comet damals sein Stand hatte/ noch etwas wolckicht / als hat der Comet=Stern / so darunter verborgen/und allem Anzeigen nach/ bereits untergangen war/nicht können observiret werden / der grosse und helle Strahl oder Schein aber / wurde biß nach 10. Uhren verspüret. Den 17. 27. Decemb. war die Lufft bey uns / neben grausamer Kälte / gantz heiter und hell / da sich dann mehr gedachter Comet Abends um 5. Uhr gantz præsentiret / unserm Gesicht nach / auf 10. Grad von unserm Horizont erhöhet. Er stund recht Sudwest. Weilen aber der zur Seiten/Sudwerts im 8. gr. Pisc. stehende Mond/ mit seinem Licht die nechstbefindliche Stern/ziemlich verkleinert/und aus dem Gesicht genommen/als konten auch/die/ dem Cometen nechst=stehende Sterne/nicht wol gesehen werden / allem Anzeigen nach aber / stund er unter dem Æquatore oder Linea, unter dem Gestirn deß Adlers bey oder in Ganymede; und ist der erste Stern in collo Aquilæ, bey der ersten observation mitten in deß Cometen Strahl oder Schein/ Lucida Aquilæ in eductione colli aber/nechst an demselbigen / und etwas bessers hinauf Sagitta gestanden/stund also der Comet über dem TropicoCapricorni. Der schröcklich breit=und grosse Schein oder Strahl richtete sich gerad theils an / theils in Galaxia, gegen Cassiopeia, welches Gestirn er auch bey nahe erreichte. Der Comet=Stern kompt einem Stellæ fixæ secundæ, oder mehr tertiæ magnitudinis gleich / dessen Farb alle Abend etwas feuriger worden / gantz aber / und mit seinem Schein oder durchscheinenden Sonnen=Strahl/ war er so groß und entsetzlich anzusehen/als iemals einer zu Gesicht mag kommen seyn. Er gieng unter gleich nach 6. Uhren / der Schein aber / wurde noch etliche Stunden / und fast biß 11. Uhr gespüret. Den 18. 28. Decemb. ist der Comet abermal um 5. Uhr Abends gesehen / und in der Erhöhung um unsern Horizont unserm Gesicht nach 13. gr. 30. min observiret worden / gieng unter um 7. Uhr. Sein Lauff ist sehr schnell/also/ daß er proprio motu in Tag und Nacht über 3. gr. sich weiter herauff über den Æquatorem den Signis Borealibus nähert / und bereits über Aquilam, Sagittam, Delphinum, Equum minorem, in Pegasum gestiegen / machet mit dem Schein oder Strahl und der Galaxia ein Angulum acutum,

so sich zwischen Andromeda und Cassiopeia zuspitzet; wie nach genugsam eingenommenen Observationen , so lang er wird können bey uns gesehen werden / in einem Astronomischen Abriß soll gewisen werden.

Was nun diser Comet bedeutet / oder nach etlicher Meynung / würcke / darvon ist von vilen/ auch viles sagen / schreiben / und urtheilens / und will iedlicher vil wissen ; da dann manchmal (auch bey denen / die vermeinen daß Hirn am rechten Ort zu haben /) so seltzame Schnacken und Mucken fliegen / daß man nicht weißt / ob man im lesen / und hören / einen lachenden Democritum, oder weinenden Heraclitum abgeben solle. Gleich wie aber noch kein Mathematicus oder Physicus in solche Experienz kommen / daß er eigendlich / was ein Comet seinem Wesen nach seye / aus gewisen Principiis darthun könte ! also auch / ob schon von gelehrten Theologis und Mathematicis, durch lange Erfahrung (worbey wir billich auch verbleiben) observiret worden / daß solche Cometen gemeiniglich grosse Aenderung / meistens aber / Unglück und Straffen vorbedeutet ; So kan doch mit keiner Gewißheit / oder einigem principio dargethan werden / was / oder wem / oder wann/ welchem Land / Nation oder Persohn / etwas hiermit vorbedeutet werde : Derowegen sich in dergleichen herauß lassen wollen / heißt eine unverständige und ungegründete Vermessenheit / und Eingriff in die Göttliche Providenz begehen.

Sehr gut/und von grosser Wichtigkeit seyn hiervon die Wort deß niemahl genug belobten ersten Teutschen Kaysers Caroli Magni, über den Cometen so zu seiner Zeit Anno 814. gesehen worden : Timeamus conditorem Cometæ, non Cometam, & laudemus clementiam ejus, qui nostram inertiam, cum peccatores sumus, talibus dignatur admonere prodigiis. Laßt uns den Schöpffer deß Cometen scheuen und förchten / und nicht den Cometen an sich selbsten ; Laßt uns seine Güte und Barmhertzigkeit preisen/ daß er um unserer sündlichen Faul=und Trägheit / durch solche Lufft=und Wunder=Zeichen/uns erwecken und aufmunteren will. Der Allweisse GOtt/aller Stern und Cometen HErr und Schöpffer / verleyhe gnädiglich / daß durch dise grosse Lufft=und Cometen=Fackel/vil in Sünden verfinsterte Hertzen / zur wahren Busse geleitet / und seeliglich / die vermuthet vorstehende Plagen und Straffen / entweder gäntzlich von uns abgewendet / oder doch gelindert werden.

    -        - Si quid scis rectius istis,
    Candidus imperti; si non, his utere mecum.

f.

T. N.

Gedruckt zu Augspurg / bey Jacob Koppmayer.

The Great Comet of 1680 over Weiltingen in south-central Franconia. The observer in the foreground is clearly measuring the tail length with a quadrant. *Anonymous*

COMET welcher Anno 1680 vnd 1681 beobachtet worden.

AEQUATOR

AEQUATOR

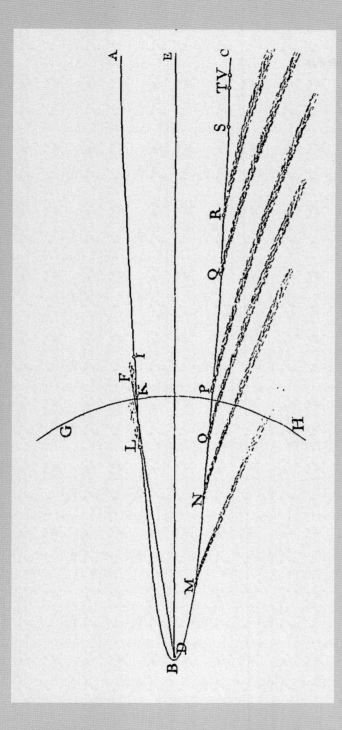

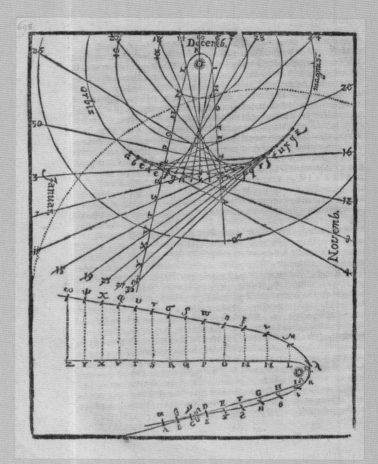

▲ Both Newton (*above*) and Dörffel (*above right*) determined a parabolic orbit for the comet.

▶ Although more and more authors raised their voice against the fear of comets, Christian authors were also active. Title page of a tract against 'Comet ridicule' by *Erasmus Finx*.

◀ View of the comet over a snow-covered winter landscape (*opposite, above*) and a stylized celestial chart with the path of the comet (*opposite, below*). The second, brighter comet at the top is an enlarged reproduction.
L. Doomer (opposite, above), M. Merian (opposite, below)

Verwerffung
Des
Cometen-Gespötts/
Oder
Gründliche Erörterung
der Frage:
Ob der Comet ein/ oder kein Straff-Zeichen
sey: Etwas oder nichts/ gutes oder böses
bedeute:
Worinnen die Vor-Bedeutlichkeit mit unverwerff-
lichen Beweißthümern begründet wird/
Die Ungründe aber der Widersprecher klärlich
entdeckt;
Neben dem auch die vielfältige Gedancken/ von dem
Ursprunge des Cometens /mit eingeführet
werden.
Auf Veranlassung des neulichst-entstandenen Wun-
der-grossen und unvergleichlichen
Comet-Sterns/
Zur Abwarnung von roher Sicherheit / und Beförderung
Christlicher Bußfertigkeit herausgegeben/
Durch
Theophilum Anti-Scepticum,
sonst zu Teutsch
Gottlieb Unverrucht
genannt.

Gedruckt im Jahr 1681.

☄ Comet Halley 1682

After the turmoil in 1680, the comet of 1682 initially aroused little attention. Only two decades later did the significance of this comet become clear. The young Englishman, Edmond Halley, who was himself one of its observers, recognized that the comets of 1531, 1607 and 1682 were actually the same, periodically returning, celestial body. The aura of being a harbinger of fate was removed from comets. In doing so, enlightenment and science gained a victory over superstition and the fear of comets.

Data	
Number:	9
Designation:	1P/Halley
Old designation:	1682 I
Discovery date:	23 Aug 1682
Discoverer:	Unknown
Perihelion date:	15 Sep 1682
Perihelion distance:	0.5826 AU
Closest Earth approach:	31 Aug 1682
Minimum Earth distance:	0.4231 AU
Maximum magnitude:	0.5
Maximum tail length:	30°
Longitude of perihelion:	109.2°
Longitude of ascending node:	54.9°
Orbital inclination:	162.3°
Eccentricity:	0.9679230

All dates are in the Gregorian calendar.

Orbit and visibility

Halley's Comet in 1682 also traced its path through northern starfields. At closest approach to Earth and perihelion the comet was north and east, respectively, of the Sun. In this respect, European and North American observers were favoured. Only the end of visibility occurred in constellations that were reasonable for southern observers.

At its discovery the comet was east of Auriga. It was already visible well away from sunrise in the morning sky. Because its path initially took it in a north-easterly direction, for Central European observers it was circumpolar between 20 and 30 August. The most favourable observational conditions occurred in the evening and morning. At the end of this phase, closest approach to Earth occurred on 31 August.

At the beginning of September, the comet moved back towards the south. It was now visible only in the evening sky. On 5 September it passed the constellation of Coma Berenices, and a few days later pas-

sed to the south of Arcturus. In subsequent days, visibility conditions deteriorated, because it was now nearing the Sun. Around 20 September it became invisible for observers in Central Europe.

The comet crossed the ecliptic well ahead of the Sun at the beginning of October, in the constellation of Libra. It was no longer detectable with contemporary methods, because it had travelled too far away from the Earth.

Discovery and observations

No unambiguous information about the discoverer of the comet has survived. North American sources favour Arthur Storer, the first astronomer in the English colonies in America, who saw the comet on 23 August. But there are sightings from France and England on the same date. Because the comet at that time was already about magnitude 2 in brightness, and could be easily seen with the naked eye, we can assume that it was noted simultaneously by astronomers in several countries.

On 24 August, Johannes Hevelius observed the comet from Danzig, and detected a tail 12° long. The following evening, from Plauen, Georg Samuel Dörffel estimated the magnitude at 1. Hevelius confirmed the tail length that he had previously reported on 27 August, and on 28 August, at the Paris Observatory, as much as 30° was seen. But John Flamsteed at Greenwich reported only 5° on 30 August, but 10° on 1 September. Hevelius estimated it at 16° on 31 August.

The widely varying values are probably not the result of morphological processes in the comet, but are to be ascribed to differing observational conditions. The most reliable are probably the sketches by Isaac Newton, according to which the length of the tail between 29 August and 1 September was about 20–25°. The comet must have then been about magnitude 1.

About 5 September, it reached greatest magnitude at about 0.5. On that date, the person after whom the comet was later named, observed it. To Edmond Halley, when compared with the Great Comet of 1680, the head appeared larger, but the tail was significantly shorter. On 8 September, Hevelius telescopically observed a curved, bright streamer from the nucleus, and similar observations were made by Robert Hooke. These were the first records of sudden outbursts that were later found to be typical phenomena of the comet.

On 12 and 13 September, the tail had already almost disappeared in the bright twilight. Halley was able to follow the comet until 20 September. On that date the magnitude must have been about 3. Later sightings were recorded on 22 September by Dörffel and Storer, and from Greenwich. The last observation was obtained by Simon van der Stel, the first Governor of the Dutch colony at the Cape of Good Hope, who, on 24 September 1682, was the last person to see Halley's Comet in the seventeenth century.

Background and public reaction

Just two years after the most impressive comet of the seventeenth century, the last 'unannounced' apparition of Halley's Comet was overshadowed by that event. At the same time, astronomers' anticipation of another bright comet was very high. Practically all prominent scientists of the time participated in observations. Johannes Hevelius, who had already become famous with his Cometographia of 1668, made observations despite his observatory being destroyed by fire three years earlier.

Among the observers was a certain Edmond Halley. The young, 26-year-old astronomer had married shortly before. The comet of 1682 was, however, to play a greater role in his life at a later date. In 1695, Halley calculated the orbits of 24 historic comets. In doing so, he was struck by the similarity between the orbits of the comets of 1531, 1607 and 1682. For a long time he doubted the agreement, but eventually, in 1705, published the prediction that the comet would reappear in 1758. The differing intervals he rightly explained as caused by perturbations

by the planets. Halley additionally predicted the return of the comet of 1680 after 575 years – but this error remains a minor footnote to his epoch-making achievement.

Among the general populace, the comet of 1682 did not meet with a great response. Only a few broadsheets appeared. However, the Turks, in their siege of Vienna, quite deliberately used the fear of comets against the defenders. Criticism of the astrological and Christian interpretation of comets was more common. In the cause of enlightenment, the scientist Jacob Bernoulli, from Basel, castigated comet mania and its pamphlets. Comets were not a sign from God. In 1682, Bernoulli wrote that one could predict as accurately "that the frogs in the Rhine would croak more often". In fact, eventually, Halley's successful orbital analysis was the fatal blow for comet hysteria. Far from being unpredictable signs, comets became predictable celestial bodies.

In many sources about the appearance of Halley's Comet in 1682 there is confusion about the date of observations. This is because of the contemporary change between the Gregorian and Julian calendars. Whereas most of the Catholic German states, Poland, France and Spain introduced the calendar reform carried out by Pope Gregory XIII from 1582, the evangelical areas of Germany as well as England and its colonies retained the Julian calendar into the eighteenth century. In 1682 the difference amounted to 10 days. Failure to take the conversion into account occurs with the date of 15 August that is often given for the date of discovery (25 August in the Gregorian calendar, which is thus later than the actual date of discovery of 23 August), and equally over the question of whether Storer or Dörffel should be awarded the honour of discovery. Both observers were using the Julian calendar, and so were 10 days 'ahead of' their contemporaries.

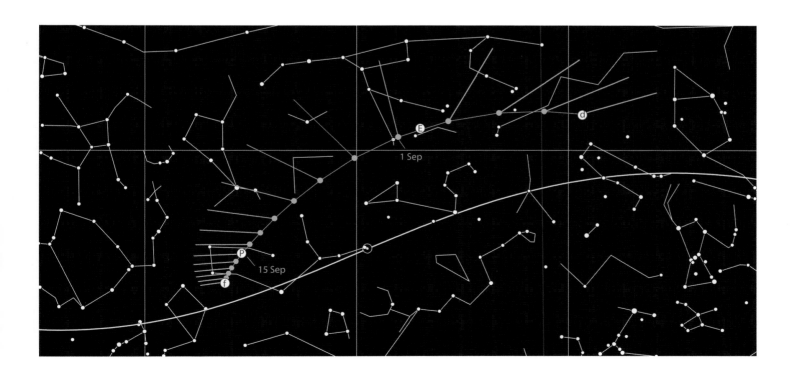

Abbildung

Deß jenigen Cometen/ welcher in diesem 1682. Jahr erschienen/ und sich zu Leopold-Stadt in Ungarn vom 10. Februarii etliche Tage nacheinander sehen lassen; sampt beygefügter außführlich- und warhafften Beschreibung desselben.

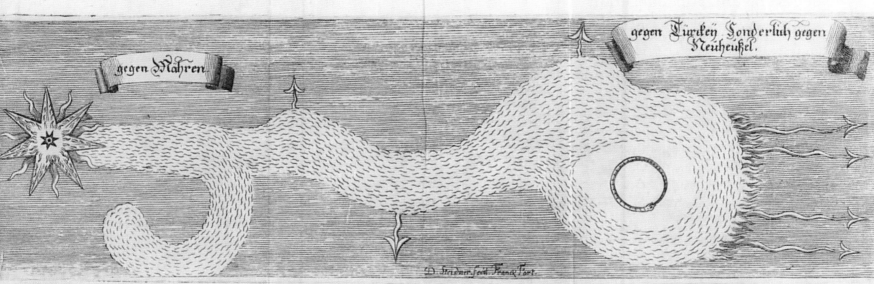

gegen Mähren

gegen Türken/ sonderlich gegen Neuheußel.

D. Steidner. fecit. Franck. Port.

ES hat der barmhertzige / zugleich aber auch gerechte GOtt je und allweg im Gebrauch gehabt/ wann er die Boßheit der Menschen straffen wollen / dieselbe zuvor entweder durch treueysserige Lehrer und Prediger / oder wol gar durch einige Zeichen am Himmel/ und in der Lufft zu warnen/ damit sie von ihrem sündlichen Leben und Wandel ablassen/ durch wahre/ rechtschaffene und unheuchlerische Busse demselben in seine Zorn-Ruthe fallen / und die angedrohete Straff von ihnen abwenden möchte. Und wird ausser allem Zweiffel kein Mensch zu finden seyn/ deme der erschröckliche Comet/ so sich im December / Anno 1680. und zu Anfang deß 1681. Jahrs in Teutschland sehen lassen/ und sich nachgehends fast durch alle vier Theile der Welt außgebreitet hat / nicht annoch in frischer Gedächtnuß schweben solte. Ob nun wol einige darfür halten / und die Meynung behaupten wollen/ (mit denen man sich aber in keinen Disputat einzulassen begehret/) daß dergleichen ungewöhnliche Schwantz-Stern und andere Lufft-Zeichen ihre natürliche Ursachen haben/ und weder Böses noch Gutes bedeuten/ so ist doch jederzeit für ein ungezweiffeltes und unwidersprechliches Axioma und Lehr-Satz gehalten worden/

quod Deus & Natura nil faciant frustra; daß nemlich GOtt und die Natur nichts umbsonst thun / und geschehen lassen / und hat die Erfahrung zu allen und jeden Zeiten viel ein anders gelehret / also / daß man die schädliche Würckungen/ solcher Cometen und Lufft-Zeichen/ so darauff erfolget sind / mehr als zu viel empfunden / wie solches mit vielen und unwidersprechlichen Exempeln darge= than und erwiesen werden könte/ wann es die Zeit und Gelegenheit zulassen solte; und ist GOtt allein bekandt / was sich noch ins künfftige auff den letzt erschienen Cometen in der Welt beydes im Geist- als weltlichem Wesen zutragen und begeben werde. Daß aber der sonst langmüthige GOtt noch nicht besänfftiget / sondern sein Zorn über uns annoch ergrimmet sey/ erscheinet darauß / indem er uns in diesem Monat ein neues Zorn-Zeichen am Himmel auffgesteckt / und in Ungarn einen neuen grausamen Cometen erscheinen lassen/ welcher zu Leopold-Stadt vom 10. Februar. an/ etliche Tage nach einander ist gesehen worden. Dieser hatte seinen Ursprung und Auffgang von der Türckey auß / und gieng auch wieder in Türckey unter. Er stunde über dem Galeociensischem Gebürg/ gieng umb vier Uhr deß Morgens auff / und verschwande mit

Auffgang der Sonnen umb sieben Uhr. Sein Schweiff und die mittlere Krümme erstreckte sich / unserm Gesicht nach/ wol in die zwantzig Elen lang / wie auß obenstehender Abbildung / welche uns von guter Hand communicirt und zugesendet worden/ eygendlich zu sehen ist. Was nun dieser neue Comet verkündigen möchte / ist zwar keinem Menschen bewußt / allein kan man hierauß unschwer abnehmen / daß er nicht viel Gutes bedeuten werde / und weil derselbe seinen Auff- und Niedergang auß und nach der Türckey gehabt / so ist wol zu besorgen / daß der erzörnte GOtt diesen grimmigen blutdürstigen Tyrannen und Erbfeind Christlichen Namens über uns schicken / und durch ihn/ die ohne dem her vorscheinende grosse Gefahr vermehren lassen dörffte/ welche wolverdiente Straff er umb seines Namens Ehre willen in Gnaden abwenden wolle.

▲ A view of the comet over Hungary. This representation reflects the fear of the Turks during the last siege of Vienna. *Anonymous*

▶ Broadsheet with the position of the comet about 25 Aug. 1682 (15 August in the Julian calendar), about a week before closest approach to Earth. *Anonymous*

Wider neu-scheinende entsetzliche

Zorn-und Wunder-Ruthe Gottes

Oder:

Cometen-Fackel/

Welche zu Nürnberg allhier ♂ den 15. Augusti Alter Zeit/und so folgend beyde Tage hernach/in diesem mit GOtt lauffenden 1682. Jahr/ mit Erstaunen vieler Leuthe sich erwiesen und sehen
lassen.

Darvon ein gelehrtes und hochreiffes Sentiment/folgendermassen gegeben worden:

Locus primæ, Phœnomeni hujus; apparitionis, prout & observatio ruditer colligi poterat, fuit in 18. long. ♌ latit. bor. 20. Hinc inter Stellas in Horizonte nostro Norico nunquam occidentes deprehensus, sub pedib. an-
terioribus Ursæ majoris. Discus hujus Cometæ, stellas primæ magn. facilè superat, nucleo claro, ovali figu-
râ inclinato, capillitio undique nimis latè circumsparso, caudam præferens satis latam, longitudinis ultrà 6. gr.

Dessen Gestaltung und Stand/ wie er Erstesmal beobachtet worden/ machet gegenwärtige
Kupffer-Bildnüs darstellig.

ES ist kein Wunder/ daß der Allmächtige GOTT
abermalen der Welt einen neuen Schrecken-Boten/
oder Cometen-Fackel an dem klaren Himmel/ als eine
Drau-Ruthe aufgestecket/ und zwar dem Sternen-Stand
nach/ zwischen dem Zeichen der Zwillinge/ und den vördern
Füssen/ des sogenannten grossen Beern. Alldieweilen man
so gar sicher und unbußfärtig in den Tage hinein lebet/ und
die leidige Straff-und Seuchen-Wirckung/ (welche nach Erscheinung deß so erschröcklichen und entsetzlichen Cometen vor
einigen Jahren/ sich aller Orten in unsern lieben Teutschen
Vatterlande hin und wieder ereignet) so gar schlecht uñ leicht-
sinnig von vielen beobachtet und behertziget worden. Dann
gewißlichen/ daß solche aufgehende rare Sternen-Liechter/
rechte Drau-Ruthen GOttes seyn/ haben von allen Zeiten
die darauf leidige Erfolgungen genugsam erwiesen und bekräf-
tiget. Und ist dahero gantz leicht und unschwer/ aus dem so
bösen und verkehrten Wandel/ der meisten Sünden-sicheren
rohen Welt-Hertzen/ in allen Orten und Enden/ zuschliessen
und abzumercken: Daß eben auch gegenwärtiger Stella Cri-
nita oder Krause Feuer-Ruthe/ uns nicht viel Gutes mit-
bringen und anzeigen werde. Es erzeiget sich aber solcher seiner Gestalt nach/ (das Corpus selbst betreffend/) als ein
ovalichter oder abländlicht eingebogener heller Feuer-Glim-
pen/welcher rings herum weit ausschweiffig/gleichsam haari-
ge Zäsern oder einen duncklen Schein von sich giebet/ und einen zimlich langen Schweiff/ dem erstmaligen erschienenen
Augen-Mas und wahrgenommener Länge nach/ bey Sechs
Graden erstreckend vor sich herführet; zu dessen Ende/auch ein

klein dunckles Sternlein annoch in dem Schweiff sich ereignet
und sehen lässet. Wolte GOtt aber/daß diese gleichsam
gedoppelte und dicke Feuer-Ruthe/ uns nicht mit ihrem Stand
unweit des Beeren/unsere rohe/ wilde/ unbändige und offt-
mals Viehische Sitten und Lebens-Wandel mit andeutete
und zu verstehen gebe. Welche so gezuckte und erhebte Straff-
Ruthe dann/ der allgüte und gnädige GOtt/ nochwohl
wieder zu ruck ziehen/und eingestellet lassen würde/ wann wir
als böse Kinder / ihme bey Zeiten in die Ruthe fielen /und
durch Buß-Bässerung unseres sündlichen Lebens/ ihme solcher Gestalt einen Einhalt thäten.

Sichrer Sünder schaue hier/ abermal ein neue
Ruthen/

Leucht und drauet über dir/ so die GOttes Straff-
Hand weist;

Stemme deinen Sünden-Schwall/ mit den Buß-
und Thränen-Fluthen/

Dann so glaube daß er dir/ ein Liebreicher Vatter
heist:

Der die Ruthe ein wird ziehen/ und sich wieder
gnädig zeigen/

So muß man zur Busse fliehen/und die Straff-
Hand Gottes beugen.

Eigentliche Abbildung/ deß am Himmel widerum herfür-

leuchtenden Comet=und Wunder=Sterns/ wie solcher den 19. 29. Augusti/ dises 1682. Jahrs/ früh gegen Tag/ ist anzuschauen gewesen.

Eigentlicher Abriß deß Entsetzliche Cometstern/ welcher widerum an dem Himel leuchtet/ und den
☿ Augusti dises 1682 J. in solcher gestalt in Augspurg ist observirt worden/ was sein fernerer Lauff und Bedeu-
tung/ folget nechsten in einen absonderlichen Bericht.

On dem Ursprung der Cometen/ führen die Gelehrte unterschiedliche Meinungen. Unter welchen zimlich der jenigen scheinbar scheinet/ welche fürgeben/ daß die Cometen/ oder andere uns erscheinende Neue Sterne/ alsbald im Anfang der Schöpffung/ zugleich aus Nichts erschaffen worden/ nicht eben zu dem Ende/ daß sie immerdar scheinen/ und leuchten: sondern ihren unsichtbaren Stand/Lauff/und Bewegung/unter dem Himmel haben sollen/ biß GOtt der HErr/aus sondere erheblichen Ursachen/selbige zu Zeiten anzündet/ der Welt etwas Denckwürdiges anzuzeigen. Andere aber/ und zwar heutiges Tages/ die meisten geben für/ daß der Comet anderst nichts sey/ als eine von vilen Dünsten zusammen gesetzte Wolcken/ welche von der Sonnen erleuchtet wird/und nachdem solche nah oder fern/so wirfft dieselbige einen langen oder kurtzen Strohm oder Schweiff von sich; wiewol auch die Dünne/ oder Dicke deß Haupts oder Kerns deß Cometen / solchen nach seiner gewissen Proportion oder Masse/ verlängert und verkürtzet. Andere haben noch unterschiedliche Meinungen/ welche wir an seinem Ort gestellet seyn lassen.

Es werden der Cometen vielerley Geschlecht angemercket/ unter welchen etliche sehr schröcklich gewesen/ als A. Christi 504. wurde ein Comet gesehen/ welcher hell flackerte/ und wie ein gehörnter feuriger Drache aufgesehen/ Deßgleichen im Jahr Christi 1527. hat sich den 11. October/ früh um 4. Uhr ein schröcklicher/ sehr langer/ und feuriger Comet/ durch gantz Europam sehen lassen/und allezeit gleichsam fünffviertl Stund gebrandt/ sein Obertheil war wie ein gekrümter Arm/ welcher in der Faust ein mächtig grosses Schwert hielte/an dessen Spitze ein heller Stern stund, an beiden Seiten aber der Schärffe waren zwey tunckele Stern zusehen; von dem Stern der Spitze/ sahe man tunckle Strahlen/ in Form eines viel - härigten Schweiffes herauf gehen/ an der Seiten schossen Strahlen/ wie Spisse/ und viel kleine Schwerter von Farben/ wie Blut; dergleichen erblickte man allerhand seltzame Menschliche Angesichter/ von Haaren/Bart/ und Farben schwartzlicht/ welche sich regten/ und so greßlich bewegten/ daß etliche darüber in Kranckheit gefallen.

Deßgleichen war der jüngsthin 1681. erschienene Comet sehr schröcklich / also / daß ihrer viel sich schwere Gedancken gemachet: weilen aber der meiste Theil der Menschen noch ruchloß dahin leben: siehe! da lösset der Höchste abermal/ zu End deß Augustmonats/ dises 1682. Jahrs/ einen nicht minder entsetzlichen Comet/ und Wunder - Stern/ an dem Himmel leuchten. Diser ist hier in Augspurg/ den 19. 29. Augusti/ früh um 3. Uhr/ erstlich observirt worden/ stehend/ über der Scheer deß Krebs/ den Schweiff gegen die vordern Klauen deß grossen Bern kehrend; selbigen Tags/ schiene der Schweiff bey dritthalb Elen lang/wiewoln er noch länger gewesen/ wan ihn nicht deß Monds Licht verdunckelt hätte. Der Kern oder Stern ist groß/un schiene an disem Tag röthlich/und gleichsam mit einem weissen Flor überzogen. Die folgende Täge aber scheinet er weiß zu seyn. Sein Lauff ist wunderlich/ dann er leuchtet Abends/ nach der Sonnen Untergang/ biß gegen 10. Uhr/ gehet alsdann Nordwerts unter/ und früh gegen 2. Uhr/ an der andern Seiten/ Nord Nord Ostwerts kommet er wiederum herfür/(daher welche vermeinet/ es liessen sich zwey Cometen sehen) seiner Länge nach/ stund er disen Tag in dem Anfang deß Löwen. Weilen er nun etlich Tag vorher anderwerts gesehen worden/ als muß er seinen Anfang genommen haben/in dem Krebs. Dergleichen Comet-Stern ist erschienen An. Ch. 745. welcher eine grosse Pest angekündiget/ so gar/ daß allein zu Constantinopel/ über dreymalhunderttausend Menschen gestorben seynd. Was nun dises Comets fernerer Lauff/ und was seine vermuthliche Bedeutung/ wird nechstens in einem besondern Bericht in Druck kommen/ da dann der Leser vilerhand seltzame Fälle mit Verwunderung wird zuvernehmen haben. Der Höchste wende alles zum Besten.

Augspurg/ gedruckt bey Jacob Koppmayer.

◀ Broadsheet on the appearance of the comet over Augsburg. Engraving. *J. C. Wagner*

▲ View of the comet above Nuremberg Castle. The instruments of the Eimmart Observatory are in the foreground. *Anonymous*

☄ Great Comet of 1744

The tail of the Great Comet of 1744 on the morning of 8 Mar. 1744 above the peaks of the Alps. The comet's nucleus was then unobservable in the southern sky. *F. Chambers*

An early morning in March 1744: six giant, curved rays towered above the southern horizon. People stared at this captivating sight, because its origin, the Great Comet of 1744 had disappeared from the celestial stage a few days earlier. The rays were the tips of the giant tail, which had developed after perihelion, and reached so far north that it was sending this final farewell to Europe.

Orbit and visibility

The comet lay approximately opposite the Sun in the constellation of Triangulum when it was discovered. As such, it was visible almost throughout the night – ideal conditions for being found. At the turn of the year 1743–1744, Comet C/1743 XI moved west from Pisces in the direction of the Great Square of Pegasus, which it entered in the middle of January. Until the end of February, the comet remained favourably placed in the evening sky and could be observed to several hours before it set.

On 17–18 February 1744, it passed the star α Pegasi in a south-westerly direction heading towards the Sun. Consequently observing conditions in the evening sky worsened, but a second window of visibility in the morning sky opened after 24 February. The tail, directed towards the north, could therefore be seen significantly before the head of the comet, because its tip rose more than an hour beforehand. Closest approach to Earth came at about this time, on 26 February.

On 29 February, the comet passed northwest of the Sun, at a distance of 10.5°. As it did so, the direction of the tail swung from a northerly to a southerly orientation. In the first few days of March after solar con-

96

Data	
Number:	10
Designation:	C/1743 X1
Old designation:	1744 I
Discovery date:	29 Nov 1743
Discoverer:	Jan de Munck
Perihelion date:	1 Mar 1744
Perihelion distance:	0.2222 AU
Closest Earth approach:	26 Feb 1744
Minimum Earth distance:	0.8266 AU
Maximum magnitude:	−5
Maximum tail length:	90°
Longitude of perihelion:	151.5°
Longitude of ascending node:	49.3°
Orbital inclination:	47.1°
Eccentricity:	1.0

Discovery and observations

Three observers – all amateur astronomers – discovered the comet independently at the end of 1743. The first was the Dutchman, Jan de Munck, who sighted the comet as early as 29 November from Middelburg. On 9 December, he was followed by his compatriot, Dirk Klinkenberg. On 13 December, it was the turn of Philippe-Loys de Chéseaux in Switzerland. The last compared the brightness with M31, which, however, seemed somewhat fainter. According to this, the comet must have been about magnitude 3.5, and already visible to the naked eye. De Chéseaux gave the extent as 5'.

At the end of December, C/1743 X1 was at about magnitude 2.5, having already become significantly brighter. A short tail was visible through a telescope. This developed faster from January 1744 and had already reached a length of 6–8° by the end of the month. However, only the first 2.5° were really bright. At this period the comet had a magnitude of about 2.

In February, the rate of increase in the brightness became markedly greater, and the comet became more-and-more brilliant from one evening to the next. By the beginning of February, it has already passed the magnitude 1 mark, and by the middle of February at −1.5, it seemed brighter to observers than Sirius, the brightest star in the sky. In February, the tail formed a beautiful, almost straight ray of light, 20–25° long and 2° wide. Jacques Cassini, son of Giovanni Domenico Cassini, saw the tail on 15 February as double: the eastern branch was 7–8° long and curved, and the western was straight and reached 24°. This division of the tail had already been seen by de Chéseaux on 31 January, and it remained until the comet passed the Sun.

In the days before solar conjunction, the sight became ever more impressive. Gottfried Heinsius described the comet on 24 February as bright and reddish-yellow 'like a firebrick'. The comet's magnitude reached that of Venus, which, at −4, was also bright in the morning sky.

On 28 February, C/1743 X1 was perceptible in the daytime sky. It could be recognized at midday with the naked eye, but this required careful scrutiny of the sky. Its magnitude must thus have peaked at about −5. Telescopic observations in daylight were made for the first time at this opportunity, and these were carried out in Italy, in particular. Until 4 March, this was possible because of the great brightness of the comet's head.

junction, the comet's head, south of the Sun, was already unfavourably placed from northern latitudes, and could no longer be seen. But the tail was so long, however, that until about 10 March, for Central European observers, it towered above the morning horizon. At the same time, Venus was at its brightest as the 'Morning Star'.

The comet then moved in a southerly direction through Aquarius, then towards the east at the border of Cetus and Pictor. As it did so, its tail finally became invisible for Europeans. This remained the case until the last observations in April 1744.

On 1 March, reports began of sightings of the tail (for which the Great Comet of 1744 became famous), reaching above the morning horizon before sunrise. From Europe, a giant, 30°-wide, 'multiple' tail towered about 20° above the horizon in the morning twilight, while the head of the comet was about 20° below the horizon. The high point was reached on 8 March. As many as six rays stretched across the sky! The sides of them appeared brighter, so that some individual observers spoke of an eleven- or twelve-fold tail. On 7 March, Heinsius even mistook the phenomenon for the aurora. However, it was actually the tips of the tail with conspicuous dark striae. (Comet McNaught showed a similar appearance in 2007.) The last sighting was reported on 9 March.

Farther south, seamen were able to observe the comet as an impressive whole. When the head of the comet rose shortly before the Sun, the tail stretched across the sky as a gigantic, strongly curved arc. The length amounted to between 40° and a gigantic 90°, according to reports from various sea captains. After 22 April there were no reports of the Great Comet of 1744. The brightness must have decreased by that time.

Background and public reaction

The Great Comet of 1744 was the brightest comet of the eighteenth century. It was the last comet that was described in the old types of pamphlets and broadsheets. The reception of these tracts is, however, not to be compared with that of the Great Comet of 1680. This reflected the increasingly rational view of the starry sky and the objects within it.

The name given to the comet is confusing. In most sources it is described as 'de Chéseaux's Comet' even though the Swiss was the last of the three independent discoverers. In the literature it is often a case of 'Klinkenberg's Comet' or 'Klinkenberg–de Chéseaux'. For a long time the earlier discovery by Munck was not recognized, because it appeared in a largely unknown publication. If, however, Munck is given priority, the official designation of the comet should actually be C/1743 W1, because it was discovered in the second half of November, and not in the first half of December.

The comet became famous for its 'multiple' tail. This illusion of a tail that was split into several parts was explained only much later. Contrary to what contemporary witnesses thought, the individual rays did not have their roots in a common source. They were far more likely to have been the upper portions of rays that ran diagonally across the tail. These forms, known as striae, are formed by a sequence of jets of material that are ejected from the comet's nucleus.

A similar appearance to that seen in March 1744 occurred when C/2006 P1 (McNaught) also passed close to the Sun, and then changed on its southern side. Like that of 1744, the head of the comet was no longer visible, but the tips of the tail, consisting of numerous striae, stretched above the horizon. The appearance of Comet West in 1976 was also compared with that of the comet of 1744.

A Representation of the COMET that appear'd on Jan: y 26 & &c. 175¾ in the Evening. Taken near S.t Martins - Church in the Strand.

A. the Comet as it appear'd to the naked Eye. B as it appear'd through a Telescope. C the Persons making y.e Observation.
Published Feb.ry 22.d 174¾ according to an Act of Parliament. Price 5.

View of the comet over London on 26 Jan. 1744. Telescopes are pointing at the comet from a rooftop. Engraving. *Anonymous*

Charles Messier, who later achieved fame as the first comet hunter, saw the comet as a young man and later wrote that this had kindled his interest in astronomy. De Chéseaux maintained that it was a periodic object with an orbital period of 442 years. Nowadays, however, it is generally taken to have been on a hyperbolic orbit – the Great Comet of 1744 thus vanished, never to be seen again, into the depths of the outer Solar System.

Eigentliche vorstellung des grosen Comets Sterns wie selbiger Anno 1680. den 8. Jan: zu Zürich ist geseh worden, und welcher in die 90. Himlische grad lang gewesen. A. Caput Andromede. B. dextra Scapula. C. Mirach. D. Medio cinguli. E. extremo cinguli.

Dessgleichen der Comet stern so den 7. Febr: 1744. zu Zürich ist gesehen worden. A. Caput Andromede. B. dextra Scapula. C. Mirach im Pegasus. D. Algenib. E. Marcab.

Prospect von dem Kratz in Zürich.

Joh. Conrad Nözli delin. *D. Herrliberger excudit.*

▲ A comparison of the comets of 1680 (*left*) and 1744 (*right*). D. Herrliberger

▼ Views of the comet in January and February 1744. *J. G. Puschner*

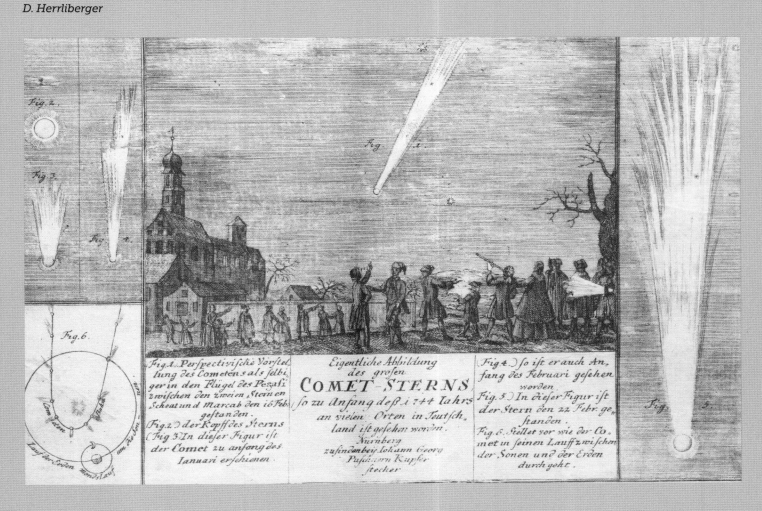

Fig. 1. Perspectivische Vorstellung des Cometens als selbiger in den Flügel des Pegasi zwischen den zweien Sternen Scheat und Marcab den 16 Feb: gestanden.
Fig. 2. der Kopf des Sterns
Fig. 3. In dieser Figur ist der Comet zu anfang des Ianuari erschienen.

Eigentliche Abbildung des grosen
COMET-STERNS
so zu Anfang deß i 744 Iahrs an vielen Orten in Teutsch land ist gesehen worden.
Nürnberg zufinden bey Iohann Georg Puschnern Kupfer stecher

Fig. 4. so ist er auch Anfang des Februari gesehen worden.
Fig. 5. In dieser Figur ist der Stern den 22 Febr: gestanden.
Fig. 6. Stellet vor wie der Comet in seinen Lauff zwischen der Sonen und der Erden durch geht.

The impression of six or more tails rising above the European horizon (*left*) gave rise to the belief that they had a common source in the nucleus of the comet. The situation actually corresponded with the appearance of Comet McNaught in 2007. The sketch (*below*) shows the path of the comet from 13 Dec. 1743 to 8 Jun. 1744. The six-fold tail is depicted for 8 Mar. 1744. *Anonymous (left), Philippe Loys de Chéseux (bottom)*

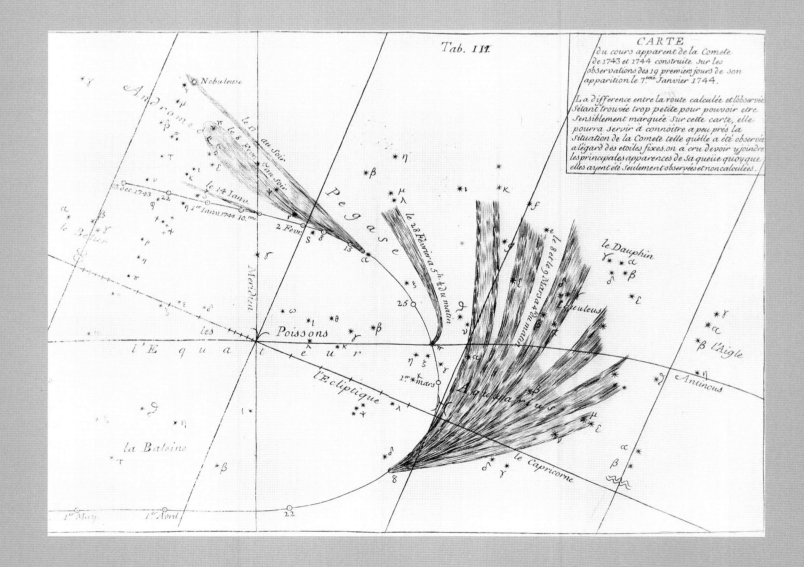

☄ Comet Halley 1759

The first return of the comet after Halley's prediction was a triumph for science. Comets were no longer unpredictable nightmares, but predictable celestial phenomena. A farmer from Saxony, of all people, played a major role in this, and his recovery of the comet was a special form of Christmas present. But in Paris people did not find this amusing.

Data	
Number:	11
Designation :	1P/Halley
Old designation :	1758 I
Discovery date:	25 Dec 1758
Discoverer:	Johann Georg Palitzsch
Perihelion date:	13 Mar 1759
Perihelion distance:	0.5845 AU
Closest Earth approach:	26 Apr 1759
Minimum Earth distance:	0.1225 AU
Maximum magnitude:	0
Maximum tail length:	47°
Longitude of perihelion:	110.7°
Longitude of ascending node:	56.5°
Orbital inclination:	162.4°
Eccentricity:	0.9676792

Orbit and visibility

The orbit of Halley's Comet at its first predicted return in 1759 was extremely similar to that of its most recent apparition in 1986. Only the beginning and end of visibility could be seen from Europe, and at closest approach to the Earth the comet was in the southern sky. Moreover, the apparition was divided into three periods of visibility for European observers, with intervening phases of invisibility.

At discovery, the comet was in the constellation of Pisces in the evening sky. At the beginning of 1759, it approached the Sun in Aquarius. The first conjunction with the Sun took place there on 1 March: because of its separation of just 7°, the comet remained unobservable.

By the end of March the comet reappeared in Capricornus. From Europe for about three weeks it was therefore unfavourably low on the horizon to be seen. Then it disappeared in the direction of the southern sky. Subsequently, the comet crossed the whole of the southern sky from Pisces Austrinus to Centaurus within 10 days, at an apparent angular velocity of up to 17° per day. On 25 April, it reached its southernmost location at –71° declination. The following day it was closest to the Earth, so it could not be followed from Europe.

In the last days of April, the comet reappeared in European skies in Hydra. When it did so, it was nearly opposite the Sun, so it could be observed in dark skies. The best period of visibility for northern latitudes then followed, although the Full Moon on 12 May interfered with the view.

At the end of May, the rapidly retreating comet reached the constellation of Sextans. Here, it remained visible until the end of June, before it disappeared in the evening twilight.

Discovery and observations

At Christmas 1758, Johann Georg Palitzsch, a farmer from Saxony, allowed himself an hour's leisure at the telescope. Using a Newtonian telescope of 8 feet focal length, that stood close to his farmhouse in Prohlis near Dresden, on the evening of 25 December he observed Mira in Cetus. Palitzsch later commented: "When, according to my laborious habit of observing everything that occurs in physics as much as possible, and to be attentive to celestial events that the fixed stars undergo and to see how the star in the Whale, which is now visible, appears, and also whether the comet, so long predicted and so greatly desired would, furthermore, be revealed, so was I granted the indescribable pleasure of discovering, not far from this wonderful star in the Whale, in the constellation of the Fish and indeed in the strip between the two stars Epsilon and Delta according to Bayer's *Uranometria*, or O and N on the Doppelmayer charts, a nebulous star never before detected in that position. Repeated viewing on the 26th and 27th confirmed my suspicion that this was a comet. Because since the 25th it had truly surged forward ... within 2 days by 3 degrees 24 seconds in longitude and 1 degree 5 seconds in latitude, and indeed in a retrograde direction."

Palitzsch reported the discovery to the Dresden scholar Christian Gotthold Hoffmann, who observed the comet on 28 December. He did publish a note about his discovery, but did not recognize that he was actually dealing with the long-sought Halley's Comet. It was only after the Leipzig astronomer, Gottfried Heinsius, had also observed the comet on 18–19 January 1759 that the identity was confirmed. Charles Messier, independently of the activity in Saxony, succeeded in finding the comet on 21 January, using a Newtonian telescope with a focal length of 4.5 feet.

In January, the comet appeared round, having a bright nucleus with a magnitude of about 6.5. No tail was visible. The comet now rapidly brightened, but because of bad weather in Europe, there were only a few observations. Messier's last observation at the very end of the first window of visibility was on 14 February, only 14 minutes after sunset, when it was about 20° away from the Sun. The next evening he did not succeed in seeing it.

On the island of Réunion, Jean Baptiste François de la Nux caught sight of the comet on 26 March. Messier saw it again from Paris on 1 April. It appeared brighter than before conjunction with the Sun. Messier compared the head of the comet with a first-magnitude star, and noted that the comet had brightened. He gave the length of the tail as an impressive 25°. But by 15 April the comet had disappeared again for Messier, with the last sighting obtained just 4° above the horizon.

The phase of maximum brightness followed, which was visible from equatorial or southern latitudes. Modern estimates put the value at magnitude 0, but unfortunately few observations have come down to us, because proper observatories did not yet exist in the southern hemisphere. According to de la Nux, the tail measured 25° on 28 April, but it was thinner than previously.

On 29 April, Messier was able to see the comet again. On 1 May he estimated the magnitude as 1, and on 3 May it was even visible with the naked eye, despite the moonlight. In the telescope, however, the tail was only 1.5° long. In contrast, de la Nux reported an incredible tail length of 47° on 5 May – darker skies and a higher elevation above the horizon favoured him over European observers.

While de la Nux still reported a tail of 19° long on 14 May, Messier was already comparing the comet with a fourth-magnitude star. By 18 May, the brightness had significantly decreased, and the comet was too faint to be seen by the naked eye. Messier's last observation dates from 3 June. João Chevalier was still able to follow it with a 7-foot Gregorian telescope until the 22 June.

Background and public reaction

The apparition of 1758–59 was the first predicted return of Halley's Comet. Halley's prediction became a touchstone and triumph of science, which finally put paid to the interpretation of comets as sensational events.

In 1695, Edmond Halley had suspected and in 1705 had proposed that the comets of 1531, 1607 and 1682 were a single object. After he had recognized that the comet had passed close to Jupiter in 1681 and had, as a result, altered its orbit, he put the return at the end of 1758 or the beginning of 1759 – although he would not live to see confirmation of this prediction.

Subsequently, there were a few new attempts at the calculation. Leonard Euler brought perihelion forward to 1757, because he believed that the period was continuously shortening. Jérôme Lalande named November 1757 as ideal for rediscovery. Alexis Claude Clairault improved on these suggestions by predicting perihelion for 13 April 1759 – only one month away from the true date – but this orbit was only made public in November 1758.

Many observers searched for the comet from 1757. Among these was the young observatory assistant, Charles Messier, who had been

given this task by his Director, Joseph-Nicholas Delisle. Messier used a chart by Delisle, which was, however, based on a date at which the discovery would be made with the naked eye. Telescopically, however, it would be possible to detect the comet before this date. However, Delisle forbade Messier to search in any other locations.

So, unwittingly, he left the way clear for the 'peasant astronomer' Johann Georg Palitzsch from Saxony. Under difficult circumstances in the middle of the Seven Year's War – Saxony was under siege from Prussia and Austria – he succeeded in the epoch-making discovery. Official astronomy, particularly French astronomy, had been duped. Messier had found the comet on 21 January, without knowing anything about the actual discovery. On the very same day, the news arrived in Paris that the comet had already been found – almost one month earlier.

Delisle prohibited Messier from announcing his discovery, because he did not believe that they were dealing with Halley's Comet. Not until 1 April was Messier able to publish his observations. By that time many astronomers, in particular those of the royal observatory (now the Observatoire de Paris), which competed with Delisle's Observatoire de la Marine, did not believe in his independent discovery. This deliberate confusion was to undermine Messier's reputation for many years.

With hindsight, Messier's search for Halley's Comet should be seen as a success. During the search, Messier had actually already come across another comet in August 1758, and which de la Nux had discovered. On 28 August, this comet passed a nebulous spot, which Messier did not know – and sure enough, he confused it with the comet. This experience caused the French astronomer to draw up a list of nebulae that might resemble comets, seen through a telescope. Nowadays, the Messier Catalogue is the most famous list of deep-sky objects and the comet hunter's most famous legacy.

Halley's Comet in 1759 was more thoroughly and scientifically documented than any previous comet. Because of the unfavourable circumstances surrounding its visibility, general attention only began in April, finally peaking in May, when the comet once again stood high in the sky over Europe. Closest attention was paid to position determinations, to improve the calculations of the celestial mechanics of the comet's orbit. Wider interest in the comet was, however, low, because at greatest brightness it was not visible in Europe.

The contestants in the race to recover Halley's Comet: Johann Georg Palitzsch, the 'peasant astronomer' from Prohlis near Dresden (*above*), and Charles Messier, the young assistant at the Observatoire de la Marine in Paris (*below*).
M. Keyl (above), N. Ansiaume (below)

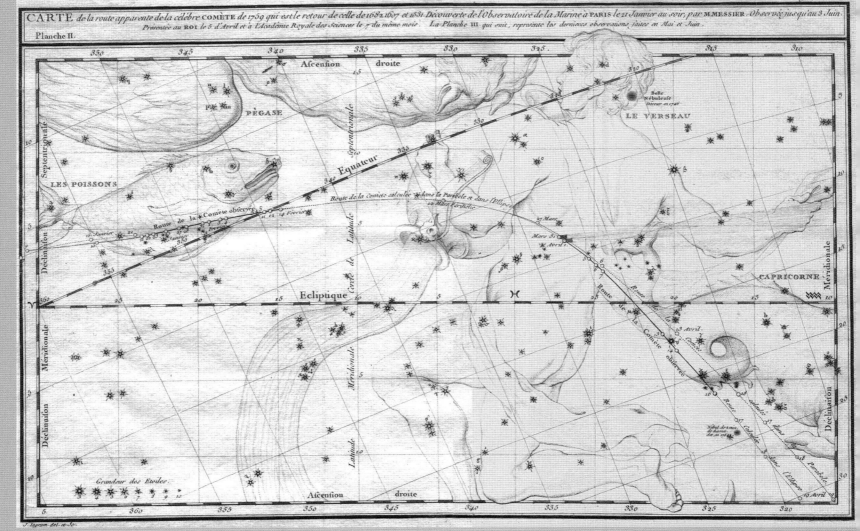

These charts show the two portions of the comet's path visible from Europe: from 16 Jan. to 19 Apr. 1759 (*above*), and from 29 Apr. to 15 Jun. 1759 (*below*).

Charles Messier

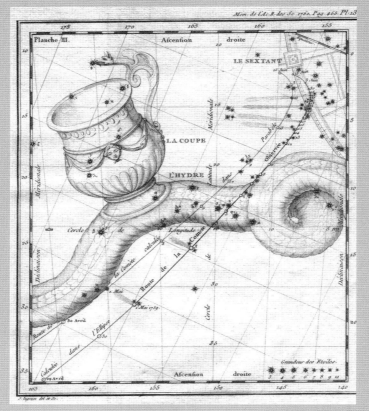

☄ Comet Messier 1769

Ten years had elapsed after Messier's defeat in the race for the recovery of Halley's Comet, when the 'ferret of comets' came up with a spectacular discovery. Messier's find rose, in the fastest time, to become the second brightest comet of the century. The length of the tail of the Great Comet of 1769 was impressive. The comet gained Messier national and international recognition.

Data	
Number:	**12**
Designation:	**C/1769 P1 Messier**
Old designation:	**1769 I**
Discovery date:	**8 Aug 1769**
Discoverer:	**Charles Messier**
Perihelion date:	**8 Oct 1769**
Perihelion distance:	**0.1228 AU**
Closest Earth approach:	**10 Sep 1769**
Minimum Earth distance:	**0.3229 AU**
Maximum magnitude:	**0**
Maximum tail length:	**60°**
Longitude of perihelion:	**329.1°**
Longitude of ascending node:	**178.3°**
Orbital inclination:	**40.7°**
Eccentricity:	**0.99924900**

Orbit and visibility

Charles Messier discovered his Great Comet in the constellation of Aries. From there it moved into Taurus in August. This meant good visibility in the morning sky with a relatively high elevation above the horizon.

In the first days of September, the comet passed through the northern area of Orion. On 5 September is lay about 3° south of Betelgeuse. Closest approach to Earth came on 10 September in Monoceros.

As it approached perihelion, the angular velocity of the comet against the sky increased notably. Simultaneously, the duration of visibility in the morning sky decreased. In the middle of September, the comet was in the constellation of Hydra, close to the Sun, and could no longer be seen.

In the early days of October, perihelion passage took place. The comet was then invisible, but passed south of it. A short time later, the second phase of visibility began in the evening sky. The comet rapidly moved away from the Sun in the constellation of Virgo and entered Serpens on 21 October, then Ophiuchus on 1 November. Here it was lost amongst the innumerable stars of the summer Milky Way.

Various authors calculated parabolic and elliptical orbits for C/1769 P1. The orbital calculation by Friedrich Wilhelm Bessel is generally accepted, and this gives an elliptical orbit with a return period of 2090 years.

Discovery and observations

The famous French astronomer, already known in specialist circles as a discoverer of comets, found the comet on 8 August 1769 with a telescope at the Observatoire de la Marine in Paris. Messier described his find as 'faint', but could, nevertheless, see the comet, later that very night, at a higher altitude in the morning sky, with just the naked eye.

Still in the month of August, Messier's Comet had grown into an impressive object. The tail, in particular, increased tremendously. On 15 August, Messier estimated it as being 6° long. Two astronomers working at the royal Observatoire de Paris, Giovanni Domenico Maraldi and César François Cassini (son of Jacques and grandson of Giovanni) determined it as already having reached 10°, and Messier, three days later, found 15°. James Cook, who was then in the Southern Ocean on his first expedition, noticed the considerable length of 42° – while the astronomers located in Paris could make out only about half this length.

The extremely narrow tail, pointing west and curving north was actually conspicuously bright only for the first half of its length. In September, Messier was also able to observe an additional increase in the length of tail. He determined tail lengths of 36° on the 3rd, 49° on the 6th, and 55° on the 9th. At the same time the coma expanded to a diameter of about 1°, and the featureless nucleus measured 4' across.

Messier's Comet reached its maximum brightness at the time of closest approach to Earth on 10 September, when it was in the neighbourhood of Procyon in Canis Minor. With its magnitude of 0 and its tail 60° long, it was a splendid sight in the morning sky. At sea, Alexandre Guy Pingré could even make it out for more than 90°, but he reported that the tail appeared so faint at the end that it partially disappeared as soon as Venus rose. Only the first 40° were bright and up to 2° wide.

Messier was able to see the comet on 16 September for the last time before solar conjunction, for Maraldi this was the 18th. To the last, tail lengths of 40° were reported. By the time of the first sightings in the evening sky on 23 and 24 October, after its reappearance, the comet was significantly fainter. Messier estimated the magnitude on 26 October as 3 and observed a tail 3° long. On 1 November he could make out a length of 6°.

On 18 November, C/1769 P1 was seen for the last time with the naked eye. The tail was by then just 2° long. The last telescopic observations by Messier were successful on 1 December. The Swedish astronomer, Pehr Wilhelm Wargentin was the last to see it on 3 December 1769.

Background and public reaction

Charles Messier was the first comet hunter. The mishap in the search for Halley's Comet 1758–59 had spurred him on to 'go for gold' with other comets. Between 1760 and 1785 he discovered twelve comets. He was the most successful discoverer of the eighteenth century and until well into the nineteenth century he was not overtaken by any other observer.

The discovery of the Great Comet of 1769 brought Messier fame. The French King Louis XV called him to a reception at which Messier was to show him a chart of the comet. On this occasion, the French king called Messier 'a cometary magpie' because no sooner had a comet 'hatched' than he had tracked it down. This led to Messier's popular nickname of 'the ferret of comets'. The increase in prestige that this brought later enabled him to marry a noblewoman.

Messier was also the most eager observer of this comet. For 42 nights he had observed it from the centre of the French capital. In 1808 there was more excitement about Messier and his comet. The astronomer, who had in the meantime reached the age of 78, dedicated it, belatedly, to Napoleon, who was born just a week after its discovery. This link by him of the astronomical object with the birth of the French emperor, was taken badly by contemporary astronomers as a backward step towards the era of astrological interpretations of comets. In doing so, Messier primarily hoped for an improvement in his then rather precarious living conditions.

The extraordinarily long lengths of this comet's tail before perihelion were astonishing. If we take the maximum value of 90° as a basis, we find an actual tail length of 3.5 astronomical units. This value cannot be real, and values of 1.5 astronomical units may be assumed. According to the Austrian comet expert, Johann Holetschek, who, at the beginning of the twentieth century, systematically researched and evaluated all cometary appearances known up to that date, the great lengths at the beginning of September 1769 were caused by the viewing geometry. The comet's tail was pointing in exactly the same direction as the comet had been moving. As a result, material remaining along the comet's path led to an optical lengthening of the tail as seen from the direction of Earth.

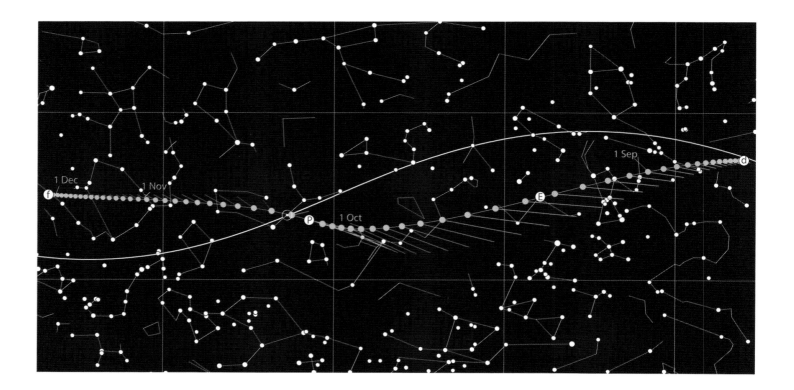

Views of the comet over Amsterdam (*above*) and over Nuremberg on 9 Sep. 1769 (*below*). *Aert Schouman (above),*
Paul Küfner (below)

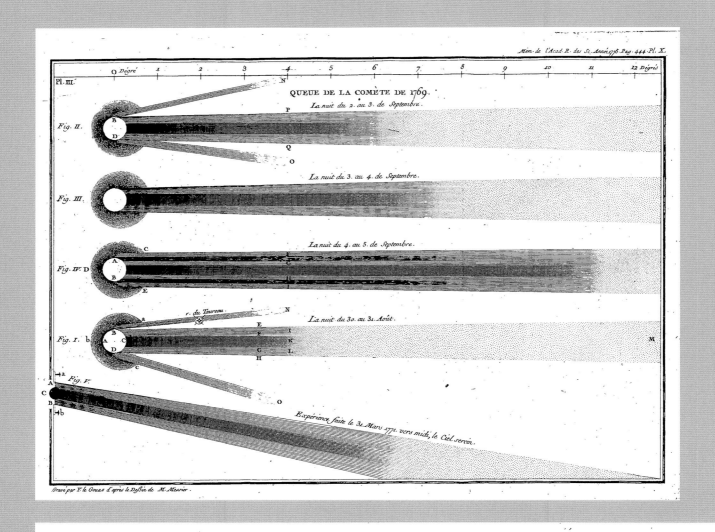

Pl. III.

O Degré 1 2 3 4 5 6 7 8 9 10 11 12 Degrés

QUEUE DE LA COMÈTE DE 1769.

Fig. II.

La nuit du 2. au 3. de Septembre.

Fig. III.

La nuit du 3. au 4. de Septembre.

Fig. IV.

La nuit du 4. au 5. de Septembre.

Fig. I.

La nuit du 30. au 31. Août.

Fig. V.

Expérience faite le 31. Mars 1771. vers midi, le Ciel serein.

Gravé par Y. le Gouaz d'après le Dessin de M. Messier.

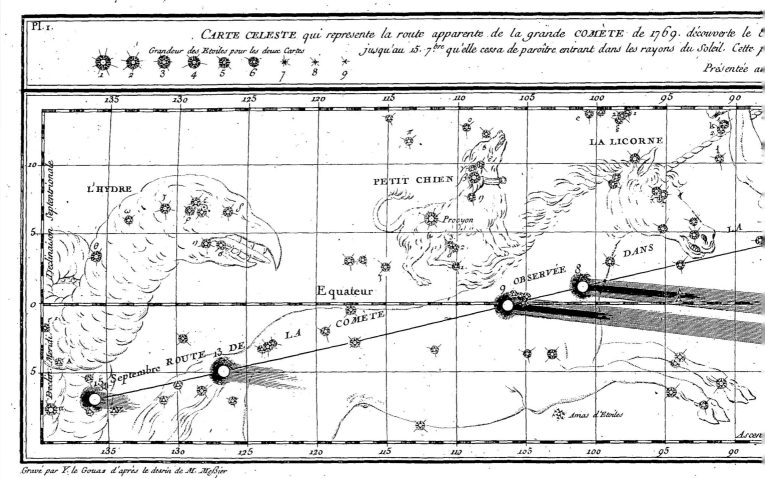

Pl. I.

CARTE CELESTE qui represente la route apparente de la grande COMÈTE de 1769. découverte le
jusqu'au 15. 7.bre qu'elle cessa de paroître entrant dans les rayons du Soleil. Cette

Présentée a

Grandeur des Etoiles pour les deux Cartes
1 2 3 4 5 6 7 8 9

135 130 125 120 115 110 105 100 95 90

LA LICORNE

L'HYDRE

PETIT CHIEN

Declinaison Septentrionale

Procyon

Equateur

OBSERVÉE

9

8

DANS

LA

COMÈTE

Septembre ROUTE DE LA

13

15

Amas d'Etoiles

Ascen

135 130 125 120 115 110 105 100 95 90

Gravé par Y. le Gouaz d'après le dessin de M. Messier.

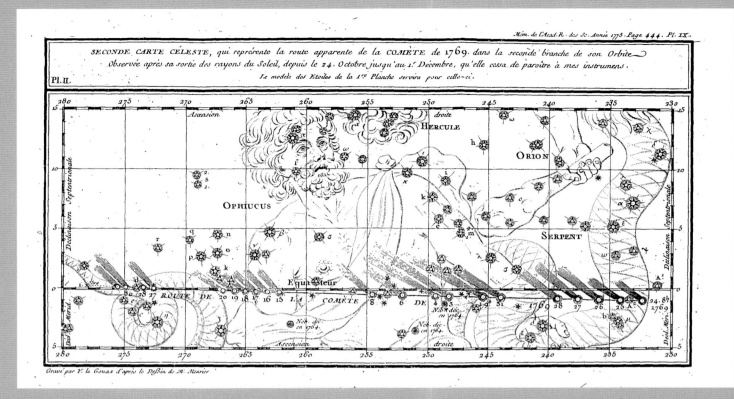

Charts and views of the Comet of 1769, drawn by its discoverer: tail structure with rays (*opposite, above*), chart of visibility before perihelion (*below*) and after perihelion (*above*). *Charles Messier*

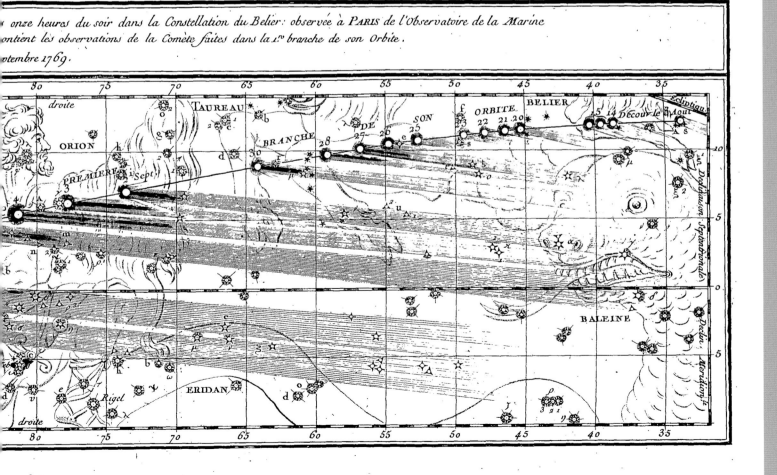

⚹ Comet Flaugergues 1811

Napoleon's rule over Europe reached its peak when the Comet of 1811 appeared in the sky. For nine months it was visible in the sky to the naked eye, although it did not come particularly close to either the Sun or the Earth. In this, its apparition resembles that of Hale-Bopp, with which it actually has a lot in common.

Comet Flaugergues on the morning of 15 Oct. 1811 from Winchester. Engraving. *H. R. Cook / A. Pether*

Orbit and visibility

The orbit of the Great Comet of 1811 is unique among the bright comets in modern times. The comet neither made a close approach to the Sun, nor did it reach a position close to the Earth. In fact, both the perihelion distance and the distance of the closest approach to Earth were more than one astronomical unit – more than any other bright comet in the last 500 years – with the exception of Hale-Bopp, the orbit of which was even farther from the Earth, but which came nearer to the Sun.

The discovery of the comet in March 1811 took place deep in the winter sky in the constellation of Puppis. Thanks to a path that took it directly north, it became ever more easily seen over the following weeks.

By the end of April it had reached the constellation of Monoceros. At the end of May, the comet passed Procyon in Canis Minor. The comet then retreated farther from the Earth to a distance of 2.41 astronomical units on 25 June. It then began to approach once more. As it did so, however, the comet became invisible in July and August, because it passed less than 10° from the Sun (in Cancer).

At the end of August the comet was again visible in the evening sky. In Central Europe it was in the northwest, near the horizon. A second observational window was available in the morning sky in the northeast. Only in September could it be seen in a dark sky, because its path took it to more northern declinations and into the southern portion of Ursa Major and into Canes Venatici. As such, it was circumpolar for a short

Data	
Number:	13
Designation:	C/1811 F1 Flaugergues
Old designation:	1811 I
Discovery date:	25 Mar 1811
Discoverer:	Honoré Flaugergues
Perihelion date:	12 Sep 1811
Perihelion distance:	1.0354 AU
Closest Earth approach:	16 Oct 1811
Minimum Earth distance:	1.2213 AU
Maximum magnitude:	0
Maximum tail length:	23°
Longitude of perihelion:	65.4°
Longitude of ascending node:	143.0°
Orbital inclination:	106.9°
Eccentricity:	0.99512500

while and thus in the sky throughout the night – just around the date of perihelion on 12 September. These favourable conditions for northern latitudes prevailed until October.

In the middle of October the comet passed through the northern area of Boötes and then into Hercules. On 31 October, it reached its maximum distance from the Sun of 67° – the comet was still visible almost the entire night. In November, the comet crossed the summer Milky Way in Aquila. In December it moved back into the evening sky.

The second pass near the Sun occurred on 17 February, bringing the best visibility of the comet to an end. The distance from our daytime star was 9.5°. Subsequently, the comet reappeared in the morning sky, but it was now at a much greater distance. During the course of the year 1812, it remained in Aquarius. Here observational conditions were ideal

in summer, when it was in opposition to the Sun – it was at this time that the last observations were made.

Discovery and observations

The French amateur astronomer Honoré Flaugergues detected telescopically a nebulous patch close to the horizon on 25 March 1811. The following night the motion of the object could be confirmed. Observations by other astronomers first came in April. The magnitude of the comet at this time was about 6.

By the middle of May, the comet was visible to the naked eye. However, because of the imminent first passage near the Sun, observational possibilities worsened. Flaugergues last saw the comet on 29 May and the Hungarian astronomer Franz Xaver von Zach on 2 June. Alexander Humboldt was last able to see it on 16 June at a separation of 40° from the Sun.

Flaugergues himself succeeded in finding the comet again on 18 August. Friedrich Wilhelm Bessel saw the comet on 23 August at an elevation of just 4° above the horizon and in twilight. He estimated the magnitude as 2. From Berlin, on the same day, Johann Elert Bode noticed a tail. It became easier to see in the succeeding days. On 9 September, William Herschel described it as curved, 10° long. On 18th, it was 12°. Herschel noted tail rays at the edges of the main tail, and compared the nebulous appearance with that of the Orion Nebula.

The comet reached its maximum brightness in the second half of October. At about magnitude 0 it was conspicuous in the night sky. The bright tail was especially impressive, which remained easily visible even at Full Moon. Herschel gave the length of the tail at 17° on 12 October, and 23° on the 15th. A characteristic of the tail was many rays, and two were particularly bright. The coma appeared about as large as the Moon, surrounding a nuclear disk of about 1' in diameter.

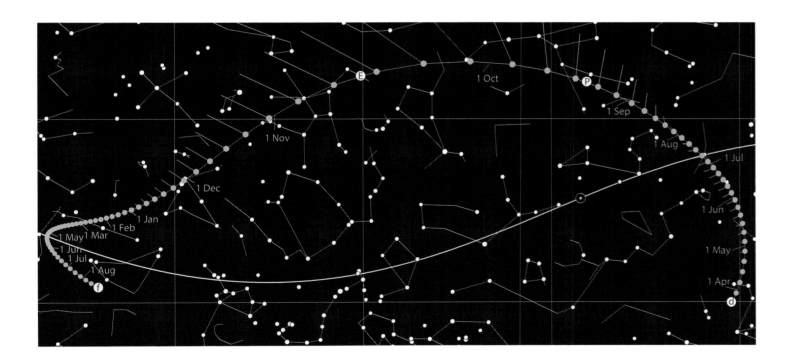

In November two components to the tail became obvious. On the 9th, Herschel reported that the brightness of the comet equalled that of the Milky Way. In December, the comet was hard to see without optical aid. The length of the tail had reduced to 7°.

Von Zach saw the comet on 11 January 1812 for the last time before the second approach to the Sun, when it was just 29° from our daytime star. The magnitude still amounted to about 5. It was found again on 11 July 1812 by José Joaquin de Ferrer on Cuba with a 4-inch refractor. The comet was then still about magnitude 8, and a tail was still present. The last sighting of the comet was obtained from Russia on 17 August, when the comet was already at a distance of 3.55 astronomical units and has already faded to magnitude 10.

Background and public reaction

Comet Flaugergues could be seen in the sky with the naked eye over a period of nearly nine months. Until the appearance of Hale-Bopp in 1997, this made it the record-holder. Overall, C/1811 F1 was followed for 511 days. As well as its long period of visibility, the orbital geometry also resembles that of Hale-Bopp. Like the latter, Comet Flaugergues was both far from the Sun and far from Earth. The great brightness, despite this, indicates an extremely large and active comet. The coma reached the gigantic size of 1.7 million kilometres and was thus bigger than the Sun, while the tail had a length of as much as 180 million kilometres.

As with Hale-Bopp, it was the long duration of visibility and the high surface brightness of the tail that aroused popular interest. Napoleon considered the comet as a good omen for his Russian campaign, which he began in June 1811. In England, mocking this view, it was noted that no comet had been seen for longer in the sky since the time of Nero. In America, the comet was seen as a harbinger of the earthquake in Missouri of 16 December 1811, and of the War of 1812 against England.

The year 1811 went down in the history of wine-growing as the 'comet year'. It was a particularly good year for vintners. The Château Lafite that year was one of the best wines ever produced. 'Comet wine' was, for a long time, a synonym for a particularly good vintage. Special bottlings when a comet appeared became a tradition dating from this time that was repeated throughout the nineteenth and twentieth centuries.

Scientific results were rather limited. Sketches by William Herschel and Johann Hieronymus Schröter survive, but a systematic series of observations was not recorded. William Smyth felt that the comet of 1811 was more interesting than Donati's Comet, more than 40 years later, in particular because of the fan-shaped tail with its edge rays. No clue can be found for the often-reported tail length of 70° in December. That would be two months after the actual peak of visibility and is thus extremely unlikely. The figure may well be a transcription error, and 7° was probably meant.

◀ The 1811 vintage is considered one of the best ever known. To commemorate the fact that the comet was seen in the sky at the time of harvest, it was known as 'comet wine'. Individual vintners took the comet as a motif for their bottling. Nowadays such bottles are worth a fortune.

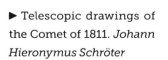

▶ Telescopic drawings of the Comet of 1811. *Johann Hieronymus Schröter*

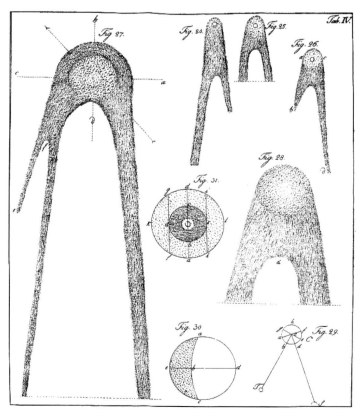

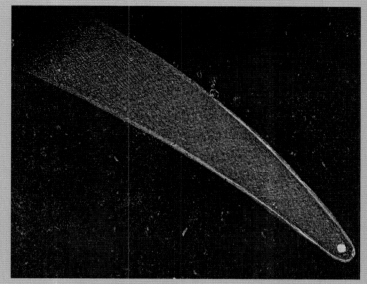

▲ A typical view of the comet. The bright edges are not an error in the representation, but one of the comet's distinguishing features. *F. Chambers*

▼ A hand-written note and sketch of the comet. *Dominikus Debler*

Different views of the Comet of 1811: over the Rhine at St Goar (*above*) and over a Silesian town (*below*). *C. H. R. Schreiber (above), R. Knötel (below)*

APPARITION DE LA FAMEUSE COMÈTE (de 1811) VUE DU QUAI DE LA VALLÉE.
Air *De la Gripette*.

Excitement over the Comet of 1811 in Paris (*above*). Another version of this scene with a different representation of the comet (*below*). *Anonymous*

▶ Drawings from 10 Sep. and 14 Oct. 1811, showing the appearance to the naked eye. The bright edges to the tail, typical with Comet Flaugergues, are prominent. *Anonymous*

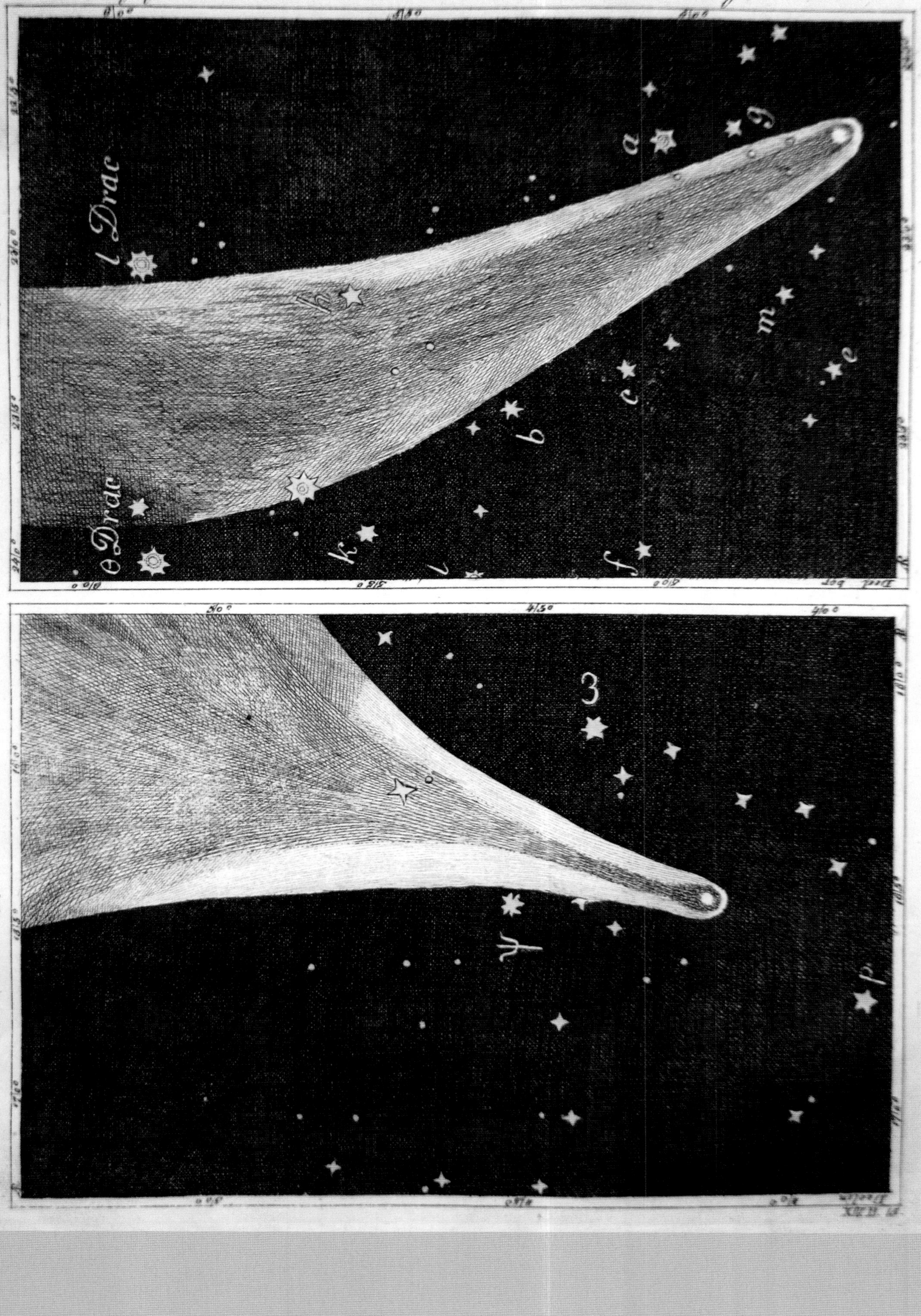
1.
2.
l Drac
θ Drac
a
g
h
m
e
c
b
k
f
ψ
3
p

☄ Comet Halley 1835

I t had been looked forward to for years: Halley's return in 1835 was another triumph for science. Free from superstitious fears, astronomers followed the comet from all over the world, and monitored it like no previous comet. The systematic observations brought new discoveries about the nature of comets. In the general public, however, it left no lasting impression, unlike comets later in the century.

Halley's Comet of 1835. Watercolours and charcoal. *John James Chalon*

Data	
Number:	**14**
Designation:	**1P/ Halley**
Old designation:	**1835 III**
Discovery date:	**5 Aug 1835**
Discoverer:	**Dominique Dumouchel**
Perihelion date:	**16 Nov 1835**
Perihelion distance:	**0.5865 AU**
Closest Earth approach:	**12 Oct 1835**
Minimum Earth distance:	**0.1865 AU**
Maximum magnitude:	**1**
Maximum tail length:	**40°**
Longitude of perihelion:	**110.7°**
Longitude of ascending node:	**56.8°**
Orbital inclination:	**162.3°**
Eccentricity:	**0.9673860**

Orbit and visibility

In the sole apparition of Halley's Comet in the nineteenth century, the visibility conditions particularly favoured the Earth's northern hemisphere. At the time of closest approach to the Earth, the comet was far to the north. This took place more than a month before perihelion, in contrast to 1759, 1910 and 1986, when closest approach followed perihelion.

When the comet was discovered on 5 August, it was close to the solsticial point on the ecliptic on the borders of Taurus and Gemini. It

had passed close to the Sun earlier, on 2 June. The comet moved away from this event in a north-easterly direction, and was thus in the morning sky when it was discovered. Conditions improved greatly for European observers in the following weeks. In September the comet passed between Boötes and Gemini, and then made its way across Lynx in the direction of Ursa Major. At the end of September it had become a circumpolar object for most of Europe.

Closest approach to Earth occurred when the comet was in the 'front paws' of Ursa Major on 8 October. A few days later, the comet reached its greatest northern declination, +64° – the farthest north for all apparitions of the comet in modern times.

After it had passed through the body of Ursa Major, Comet Halley turned back towards the south. After 15 October it returned to the evening sky. The path took it past Boötes towards the Sun and as a result the visibility worsened considerably. Perihelion occurred on 16 November. The comet was then northeast of the Sun in Ophiuchus, but for Central Europe, set at the end of twilight.

After the second pass near the Sun on 7 December, the comet moved to the southern side of the Sun. When in the constellations of Scorpius and Lupus it could at first only be observed deep in the morning sky. Only at the end of March did it return northwards, passing south of the constellation of Corvus. The last observations were obtained in May, when the comet lay between Sextans and Hydra.

Discovery and observations

As in 1759, the search for the reappearance of the comet turned into a race. This one was won by the French Jesuit Dominique Dumou-

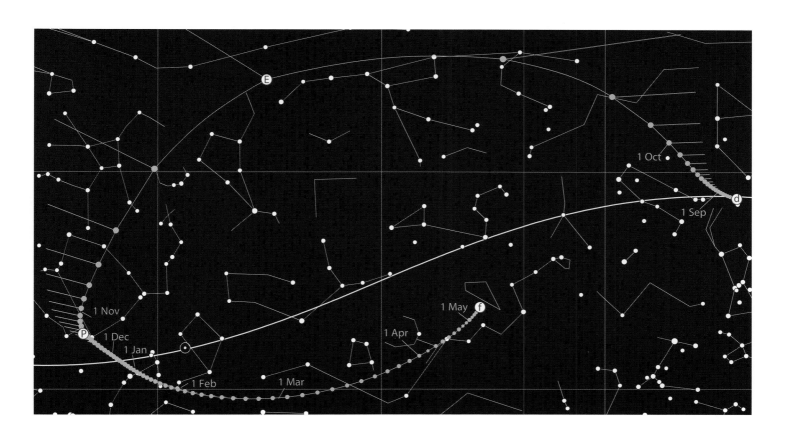

chel, who was Director of the observatory at the Collegio Romano in the Holy City. He observed the comet on 5, 6 and 7 August 1835 with the 6.5-inch refractor of the institute, which was the precursor of the present-day Vatican Observatory. On 20 August it was also found by Georg Friedrich Wilhelm Struve at Dorpat (now Tartu in Estonia) using the famous 9-inch refractor built by Fraunhofer. In the days that followed, it was also seen at other observatories.

At the time of discovery, the magnitude was about 12. When Struve saw it, it was already 10.5. However, in August the comet appeared tail-less. The first sighting with the naked eye was reported on 19 September. This was confirmed by the elder Struve on 23 September, together with his son, Otto. At that time the comet had a magnitude of about 5. One day later, the beginnings of a tail were reported, and on 2 October, Friedrich Wilhelm Bessel, observing telescopically, saw one of the jets that were to become typical for this apparition of the comet, and which he described as 'like a flame'.

The magnitude now rose significantly. On 4 October it was estimated at magnitude 4.5, and the tail grew to more than 1°. On 8 October these values had increased to 1 and 20°, although widely differing tail-lengths were reported, most likely because of differences in the clarity of the sky.

On 10 and 12 October, Struve observed a distinct jet arising from the nucleus. He compared the appearance with the 'discharge from a cannon, with the sparks swept away by the wind'. William Smyth in England and Bessel in Königsberg had noticed the jet, which, according to the latter had a length of 30" and was pointing in the opposite direction to the tail. Bessel tried to determine its rotation.

The length of the tail peaked at 40° in the days around 12 October, although this value was reached only in very dark locations. Bessel estimated it as 28° on 15 October. Smyth reported 15–20° on 19 October. The magnitude at this period was 1.5, but toward the end of the month it declined significantly. Around the end of October and the beginning of November, the comet was still at magnitude 3, and the jets from the nucleus were still visible.

The last observation before solar approach was obtained by the observatory at Kremsmünster in Austria on 22 November, only 3.5° above the horizon in the evening twilight. After passing the Sun, Halley's Comet was recovered on 30 December 1835 by Karl Kreil at the observatory in Milan. It mgnitude was about 4. Thanks to its southern location, it was also within reach of John Herschel, who was then working at the Cape of Good Hope, and who later recalled that he had 'eagerly looked out for the comet', well before he would have been able to see it.

In March 1836, a tail could no longer be seen, and the comet faded below the naked-eye limit. Herschel observed the comet "the whole night, and every night" until 20 May. "To be honest, I am glad it has disappeared" he noted in a letter written to his aunt Caroline. On the same date, a last observation was also reported from Germany. Palm Heinrich Ludwig von Boguslawski saw the comet from Breslau, at an imposing distance that was by now 2.73 astronomical units.

Background and public reaction

The second return of Halley's Comet, according to Edmond Halley's predictions, was expected in 1835. The Turin and Paris academies of science both awarded prizes for the calculation of the expected path and the date of perihelion. There were, therefore, numerous competing calculations in advance of the comet's return. Perihelion dates between 4 and 15 November were named. It was thanks to these astronomers' calculations that the comet could be found even at a distance of 2.5 astronomical units.

For the first time in the history of the observation of Halley's Comet, scientifically accurate documentation of its position was achieved. More than 1500 positional determinations were carried out by numerous observatories all round the world. The scientists' attention was also directed to the nucleus and its changing appearance: Herschel, Smyth and Bessel all tried to track its changes by accurate drawings.

Bessel's life was completely linked to Halley's Comet. As a 20-year-old, he had re-calculated the orbit for the 1607 return and, as a result, obtained his first astronomical position at Lilienthal Observatory. In 1835, Bessel recognized the link between the jets near the nucleus and the ejection of material from the sunlit side of the comet. He showed that the curved tail shape was caused by a force emanating from the Sun, that had carried cometary material away from the head, and which had previously been evaporated by solar radiation. He calculated the mass-loss as a thousandth of the total mass of the comet. From this he concluded that the next return to perihelion would occur about 1107 days early. In fact, this estimate was seriously in error – the period increases, on average, by about four days from perihelion to perihelion. But Bessel was not the only one to gain epoch-making insights: François Arago determined the polarisation of light from a comet for the first time.

The development of the coma and tail structures of Halley's Comet of 1835. Four views of the region around the nucleus, with a clearly visible jet (*top left*). Tail and coma on 29 Sep. & 3 Oct. (*top right*); 14 Oct. & 27 Oct. (*bottom left*); and 29 Oct. (*bottom right*). *John Herschel (top left), Friedrich Wilhelm Struve (top right, bottom)*

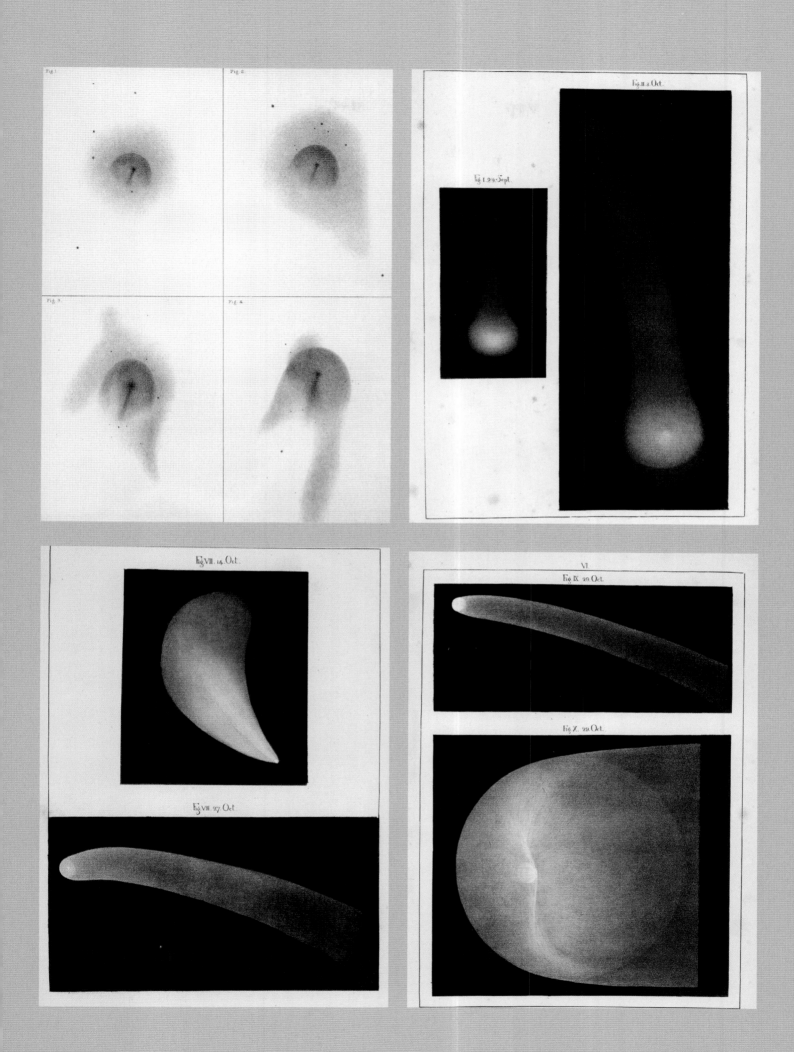

Great March Comet of 1843

It was suddenly there, near the Sun: The Great March Comet of 1843 was so bright that it was seen in the daytime sky by many people on every continent. A few days later, the giant comet reappeared in the sky, holding the record to this day for the length of its tail. Witnesses described it as a 'comet of the century'. The cause of this magnificent sight: no comet in modern times had come so close the Sun.

The Great March Comet of 1843 above the hills of Tasmania. Watercolour. *Walter Synnot*

Data	
Number:	15
Designation:	C/1843 D1
Old designation:	1843 I
Discovery date:	6 Feb 1843
Discoverer:	Unknown
Perihelion date:	27 Feb 1843
Perihelion distance:	0.0055 AU
Closest Earth approach:	6 Mar 1843
Minimum Earth distance:	0.8420 AU
Maximum magnitude:	−10
Maximum tail length:	70°
Longitude of perihelion:	82.6°
Longitude of ascending node:	3.5°
Orbital inclination:	144.4°
Eccentricity:	0.99991400

Orbit and visibility

Just 830 000 kilometres separated it from the surface of the Sun. The March Comet of 1843 was one of the most extreme sungrazers and, among the comets seen in modern times, was the one with the smallest perihelion distance. It brushed past the Sun on 27 February at a rate of 570 kilometres per second. It belongs to the Kreutz Group of comets, whose members are assumed to have originated in a parent body that fragmented in 371 BC. Kreutz himself predicted its return for AD 2356, but other calculations came out with a date 100 to 200 years later.

When the March Comet was first seen it was still February. At the time the comet was in the constellation of Pictor and was therefore unfavourably placed deep in the evening sky for Central European observers. Only at the latitude of the Mediterranean or farther south

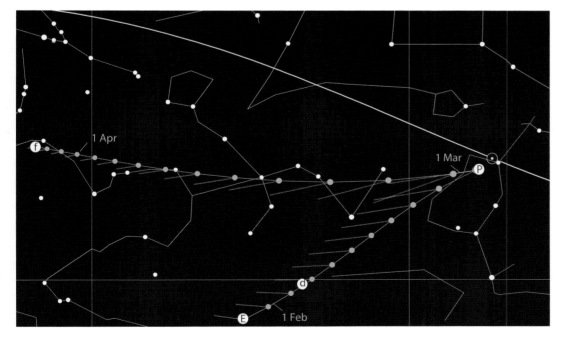

was it better seen. Northern observers had even worse luck, in that the tail lay parallel to the horizon.

The orbit did come to the aid of European observers, however, because the comet moved northwest. On 11 February it entered Cetus, and passed southwest of the bright star Deneb Kaitos in that constellation. But at the same time it approached the Sun, so the more northern location did not result in better visibility.

On 27 February, perihelion passage occurred in Aquarius. Initially, about 21:00 Universal Time (UT), it passed behind the Sun and then, just two hours later, between 23:16 and 00:29 (UT) it was out of sight for observers on Earth, as it passed in front of the Sun's disk.

The main period of visibility then began. The comet moved away from the Sun in a straight line directed towards the east. In doing so, it was better positioned for observers in the northern hemisphere than before perihelion. Because the Sun itself was also increasing in northern declination, for Central Europe, until the middle of March, the head of the comet set at the same time as the Sun.

Until 15 March, the comet was in the constellation of Cetus in the evening sky. It then moved through Eridanus and into Orion. It only reached the stars of Orion's Belt, however, at the end of May, by which time it was no longer seen.

Discovery and observations

It can no longer be determined who first saw the Great Comet of 1843. In any case, it was not an astronomer, but various inhabitants of North America and the Caribbean, whose sightings from 6 February (and 12 February, at the latest) have been preserved. At that time the comet had a magnitude of 3 to 4.

Eye-witness accounts and the amount of attention it received declined in the course of February, because the comet was approaching the Sun. Only on the day of perihelion did it return to the limelight. From Chile, Peleg Ray reported on 27 February that he could see the comet, in daylight, immediately next to the Sun, before perihelion. This was the first of extraordinarily numerous reports from people all over the world, that the comet could be seen next to the Sun, after perihelion, without any optical aid. Reports from Europe, North America, South Africa and Asia described the comet as 'brilliantly white' or 'as bright as the Moon at midnight', even though the comet was only 2–4° below the Sun. A tail of up to 5° in length could also be seen, in part, with the naked eye. Telescopically it was 10° long – and that was in the daytime sky! The

The Great March Comet on 19 Mar. 1843 above Paris. Lithograph.
Anonymous

magnitude of C/1843 D1 must have amounted to at least –10 at that time.

On 1 March, when the comet had moved far enough away from the Sun for it to be seen in bright twilight from southern locations, a very narrow tail, with the impressive length of 70° reached as far as the nighttime sky. Through a telescope, the nucleus appeared like a planetary disk, 10" in diameter.

With every day that passed, the comet, now receding from the Sun, shifted its position and gained in contrast in the evening sky. On 3 and 4 March it was a spectacular sight with two very narrow straight tails; while the gas tail measured 25° long, the dust tail reached 65°. What was striking about the appearance of the March Comet was the very tiny head when compared with the long tail. Through a telescope, the nucleus, looking like the disk of a planet, showed a golden colour, and distinct jets in the direction of the tail.

The extremely long, narrow tail remained visible over the next few days. On 9 March a length of 60° was reported. Appearing extremely bright, despite a Full Moon, the comet looked like a fine cirrus cloud – nowadays it might well be compared to a condensation trail. It was divided into two, with a very narrow interior angle, scarcely 2° wide.

Tail-lengths of 35° to 60° were observed until the middle of March. At the end of March, the comet suddenly became significantly fainter. The tail-length still amounted to more than 30°. Wilhelm Bessel thought that the comet had 'lost its power to grow any hair'. The last sighting with the naked eye dates from 1 April – the March Comet was a magnificent sight for just a month. Thomas Maclear saw the comet for the last time, telescopically, from near Cape Town.

Background and public reaction

The Great March Comet of 1843 was called the 'comet of the century' by contemporary witnesses. It was far more impressive than Comets Flaugergues and Donati. In 1849, John Herschel described it as "by far the most remarkable comet of the century". Nowadays, however, we must give equal weight to the comets of 1861 and 1882 at the very least.

The particular feature of the Comet of 1843 was the very bright and simultaneously very long and extremely narrow tail. It measured more than 350 million kilometres in length, and still retains the record for the length of a comet's tail. The naturalistic representations and watercolours by Charles Piazzi Smyth, who was working at the Cape of Good Hope when the comet appeared, are the most impressive reproductions of this apparition of a comet.

The conspicuous nature of the comet in the daytime sky was also unequalled. Whereas in most cases comets in the daytime sky require the use of optical aids or, at the very least, that the Sun should be hidden for them to be seen, the Great March Comet was so bright that it could be seen alongside the Sun. The great excitement that it caused among the general populace in Europe on 28 February, is one sign of this.

The comet also played an ignominious role. In the USA, William Miller, a farmer from New England, had predicted the end of the world and Christ's return for about the year 1843, based on extensive Bible studies. Miller had acquired a considerable following in America after 1831 from his prophecies. His followers were several tens of thousands of 'Millerites'. Miller himself avoided giving a specific date. He said that 'heavenly signs' would appear, to announce the Second Coming. Nevertheless, the date of 3 April 1843 was eventually circulated by one newspaper.

When the Great March Comet appeared, hysteria erupted among Miller's followers. They saw the comet as the sign and proof of the accuracy of his teaching. Nevertheless, April passed and nothing happened. A new date surfaced, which was accepted by most of the Millerites are the final, correct date – but the end of the world did not arrive on 22 October 1844. Despite this the day went down in history as the 'Great Disappointment'. Miller's followers dispersed, but from the remnants various Adventist groups formed, which are still active today.

▼ The Comet over Cape Town on 9 Mar. 1843 (*below, left*) and near the Sun above Table Mountain on 28 Feb. 1843 (*below, right*). Oil paintings. *Charles Piazzi Smyth*

▲ The Great March Comet during its maximum brightness, as it appeared from Europe. *Anonymous*

M M Allport.

The Great March Comet of 1843 over Aldridge Lodge in Tasmania. *Mary Morton Allport*

☄ Comet Donati 1858

No comet seems to have ever been depicted more frequently in art. Donati's Comet of 1858 unleashed a wave of comet romanticism. No more beautiful comet appeared in the whole of the nineteenth century. The fact that its great brightness was displayed away from the Sun in the night sky also contributed to this, as well as it being high in the sky over Europe. Donati's Comet was also the first comet ever to be photographed, and was thoroughly investigated scientifically.

Donati's Comet over Markree, Ireland. Lithograph. *Unknown artist*

Data

Number:	16
Designation:	C/1858 L1 Donati
Old designation:	1858 VI
Discovery date:	2 Jun 1858
Discoverer:	Giovanni Battista Donati
Perihelion date:	30 Sep 1858
Perihelion distance:	0.5785 AU
Closest Earth approach:	10 Oct 1858
Minimum Earth distance:	0.5378 AU
Maximum magnitude:	−0.5
Maximum tail length:	40°
Longitude of perihelion:	129.1°
Longitude of ascending node:	167.3°
Orbital inclination:	117.0°
Eccentricity:	0.99629500

Orbit and visibility

Donati's Comet was discovered in the evening sky. When discovered on 2 June 1858, it was in the constellation of Leo, near the head of the mythological lion. Initially, the apparent motion was small, because the comet was moving towards the Earth. It slowly moved north and then east. This caused the conditions for visibility in the evening sky to deteriorate because the Sun was continuously getting closer. The minimum distance from the Sun, at 18°, came on 10 August.

On 1 September, the comet was in the constellation of Leo Minor and shortly afterwards passed into the southern reaches of Ursa Major. In doing so, it could be seen in Central Europe both in the evening and in the morning skies. At the end of September, the apparent motion of

the comet against the sky increased. It passed across Canes Venatici in a south-easterly direction. At perihelion it was nearly 40° north of the Sun – as such it was hardly visible from the southern hemisphere, but for northern latitudes, however, this position was ideal. The comet was now primarily an evening object.

The phase of greatest brightness which then followed until closest approach to Earth on 10 October, took Donati's Comet into the constellation of Boötes. On the evening of 5 October the comet passed the principal star in Boötes, Arcturus, at a distance of just 20'. It continued on its way southeast, between Serpens, Libra and Ophiuchus. On 20 October, Donati's Comet passed Antares and Venus. The tail had changed direction and was no longer pointing north, but instead was parallel to the horizon for Europe, and directed towards the east.

This brought an end to visibility in Central Europe, but southern-hemisphere observers were favoured for the final stages. At the end of November and beginning of December, the comet passed in front of the Milky Way in Scorpius, and continued on its path southeast. The last observations were obtained in the constellation of Tucana in the southern sky.

Discovery and observations

On 2 June 1858, the Italian astronomer Giovanni Batista Donati, working at the Florence Observatory, found a small nebula about 3' in diameter. Overall, he found five comets, but this was to be his brightest find, which soon led to his name being known everywhere.

At discovery the magnitude of the comet was 7 or, according to some sources, only 10. The magnitude increased only in August, but the comet was, however, closer to the horizon than previously. When, at the end of the month, the Sun was farther from it, the first naked-eye

sighting arrived from Karl Bruhns in Berlin. The comet had a magnitude of 4 and showed a tail 0.5° long.

During the course of September, the magnitude increased significantly. At the beginning of the month it was still estimated at 3, but by the middle of September it was already 1. Over the same period, the length of the tail increased to 7°. According to Heinrich Schwabe, the tail appeared 'striped'. At the end of the month, it already measured an imposing 25°. On 29 September, the Austrian astronomer A. Reslhuber from Kremsmünster gave a description: "The tail, 20 degrees long, on its own, reached up into the night and was clearly visible when the comet was at its lowest (10° below the horizon)."

At the end of the month the magnitude was about 0. As September turned into October, six to seven shells, known as envelopes, could be seen around the comet. According to Julius Schmidt they occurred at intervals of 4.27 hours. At the beginning of October, Donati's Comet was seen in all its splendour in the night sky. The magnitude amounted to between 0 and –1. A bright, curved, dust tail stretched across the sky. From Dorpat (now Tartu, in Estonia) Johann Heinrich Mädler wrote: "The tail, which exhibited a lengthwise split from the head, so that only the outer portions formed a simple trail, reached η Ursae Majoris, was therefore 24° long." On 22 September, Mädler, in Dorpat, succeeded in seeing the comet in the daytime sky with the 9-inch Fraunhofer telescope. On 4 and 5 October, Bruhns in Berlin (using a 10-inch refractor) and George Bond at Harvard Observatory (with a 15-inch refractor) did the same. At the time, only the region around the nucleus was visible.

The comet's tail was fascinating. The length of the dust tail rose from about 25° on 1 October to 40° on 10 October. It was about 5 to 15° wide. Two parallel tail-streaks occasionally ran through it. The gas tail was considerably fainter, but up to 40° long. The English amateur astronomer William Smyth declared that it was "one of the most beautiful objects that I have ever seen". The passage by Arcturus occurred precisely during the phase of maximum brightness on 5 October. The comet then appeared somewhat brighter than the star. This encounter was captured in numerous views and drawings.

The gas tail was no longer visible after the middle of October. The striking envelopes around the nucleus vanished, and the dust tail faded visibly. Around 20 October, the magnitude was then 3.5, and the tail was still 5° long. Donati's Comet could be followed with the naked eye until November 1858. The last observation was secured by William Mann at the Cape of Good Hope on 4 March 1859.

Background and public reaction

Donati's Comet was undoubtedly the most observed comet of the nineteenth century. Its favourable orbit for Europe, in the northern sky and with a perihelion position far north of the Sun, contributed to this. The comet's great brightness was thus displayed under dark skies. In addition, widespread good weather at the peak of its visibility at the end of September and the beginning of October also played a large part.

C/1858 L1is considered to be the most beautiful comet of recent times. In particular, the classic, slightly curved dust tail and the narrow gas tail have contributed to this opinion. Donati's Comet appealed to artists enthusiastic about nature, who dedicated themselves to romantic themes. Numerous painters took the comet as their subject and depicted it in paintings and drawings.

Scientific drawings were prepared, above all, by George Bond in the USA. He documented the changeable structures around the cometary nucleus. Donati's Comet was, however, also the subject of the first cometary photograph. On 27 September, William Usherwood at Walton Common, near Reigate, Surrey, took a seven-second exposure of the nucleus with an f/2.4 lens. However, hardly anything was visible on the picture. The first telescopic comet photograph, which Bond took on 28 September, was not much better. Here the exposure was six minutes with the 15-inch refractor.

The appearance of Donati's Comet was preceded by hysteria. The noted English astronomer John Russell Hind had, from his own calculations, identified the comets of 1264 and 1556 as a single body. He predicted a return for 1857 and calculated that the comet would collide with the Earth on 13 June 1857. This led to great insecurity among the general public, particularly in Paris. It turned into a panic, and the sale of 'comet clothing' and 'comet safeguards' flourished. The confessionals in the Paris churches were full all day long. Nowadays, the comets of 1264 and 1556 are considered to be separate bodies.

Donati's Comet had a tail-length of about half the separation between the Earth and the Sun, and was thus about 75 million kilometres long. In 1979, Fred Whipple succeeded in drawing conclusions about the rotation period of the nucleus from old drawings and notes. He found a period of 4.62 hours.

Donati's Comet over the Observatory in Cambridge on 11 Oct. 1858. *W. H. Smyth*

Typical views of the Comet were widely copied and incorporated in scenes of major European cities (*left*: Florence, *below*: London and Paris). They show the Comet as it passed the star Arcturus.

▲ Numerous artists depicted the Comet in paintings. *H. Griesbach* (top), *J. Poole (centre), anonymous (bottom)*

▲ The Comet on 10 Oct. 1858. Etching. *Cornelis van der Grient*

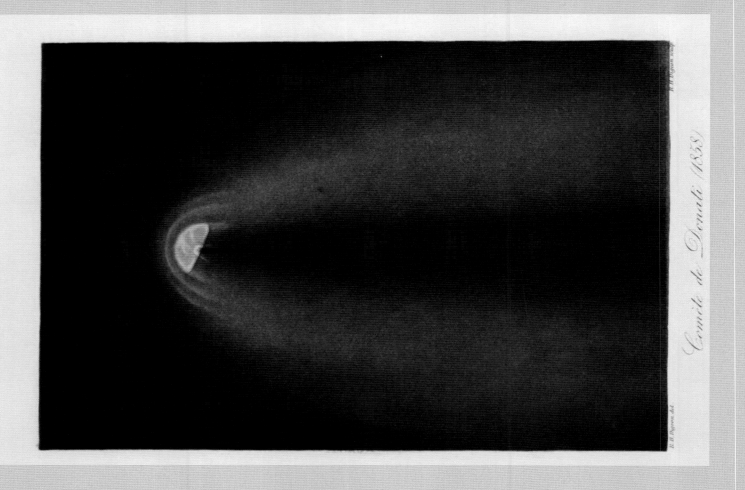

La Comète de Donati vue à l'œil nu

Comète de Donati (1858)

Comet Donati not only displayed a fascinating overall view on 4 Oct. 1858 (*opposite above*), but also impressive fine detail within its head (*opposite below* undated, *above* 2 Oct., *below* 10 Oct. 1858). *G. P. Bond*

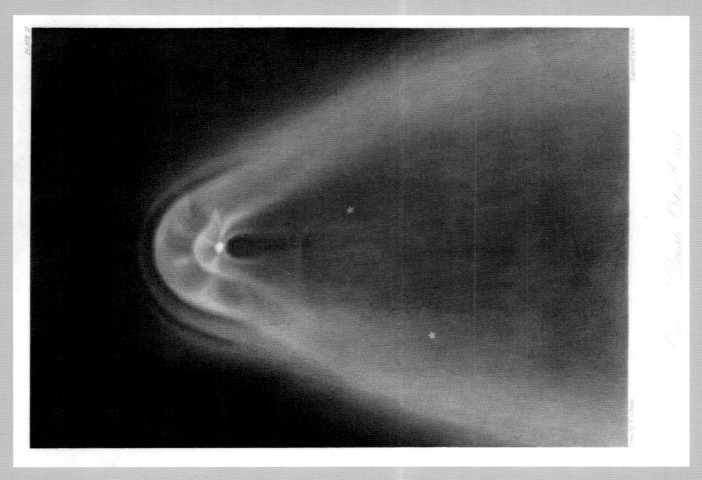

Comet Tebbutt 1861

The Great Comet of 1861, Comet Tebbutt. *E. Weiß*

Among the great comets of the nineteenth century, it was the greatest. Comet Tebbutt, discovered by an Australian amateur, was so bright that, according to eye-witness reports, it cast shadows. At the same time the tail showed the immense length of up to 120° and was extraordinarily wide and bright. Favourable orbital conditions brought the comet near to the Earth, and at the same time the Sun did not interfere with the spectacle.

Orbit and visibility

The impressive appearance of Comet Tebbutt in 1861 may primarily be linked to its close distance to the Earth. The small value of merely 19 million kilometres is a little greater than that of Comet Hyakutake in 1996. The comet's path in the sky also resembles that of Hyakutake. It lay almost at right-angles to the Earth's orbit. Closest approach to Earth occurred half a month after perihelion.

When young John Tebbutt found his Great Comet in May 1861, it was in the constellation of Eridanus, in the southern portion of the constellation, difficult to see from European latitudes. The position, south of the Sun, made it impossible to see the comet from northern

Data	
Number:	17
Designation:	C/1861 J1 Tebbutt
Old designation:	1861 II
Discovery date:	13 May 1861
Discoverer:	John Tebbutt
Perihelion date:	12 Jun 1861
Perihelion distance:	0.8224 AU
Closest Earth approach:	30 Jun 1861
Minimum Earth distance:	0.1326 AU
Maximum magnitude:	−3
Maximum tail length:	120°
Longitude of perihelion:	330.1°
Longitude of ascending node:	280.9°
Orbital inclination:	85.4°
Eccentricity:	0.98507000

latitudes, and it remained a purely southern object until shortly before closest approach to Earth.

The comet was initially visible in the morning sky. In June, its motion accelerated considerably, and it rapidly moved north in the sky. At the end of June, its path took it west of Orion in Taurus. In doing so, it first became visible in Europe and North America on 29 June. It remained west of the Sun, so was visible in the morning sky.

On 30 June, the comet reached the closest point to Earth on its orbit. It was then in Auriga, north of the Sun, and was visible throughout the night to viewers north of latitude 50° north. At the same time the Earth passed through the tail of the comet, which was 50 million kilometres long.

In subsequent days the comet passed farther north into the constellation of Lynx. In the first week of July, C/1861 J1 passed over the top of Ursa Major. As it did so, the tail swung round, on 11 July, from a westerly direction to an easterly one. At the end of July the comet crossed into the constellation of Boötes. Because it was no longer at high northern declinations, it returned to the evening sky.

In November 1861, Comet Tebbutt was located in the constellation of Hercules, by March 1862 it was in Cepheus, where it was last seen. The comet's period was, according to Kreuz, 409 years. The Japanese comet researchers Ichiro Hasegawa and Syuichi Nakano identified Tebbutt with a comet seen in 1500, and they have predicted a return in 2265.

Discovery and observations

C/1861 J1 was discovered on 13 May 1861 by the 26-year-old Australian amateur astronomer, John Tebbutt, observing near Windsor just outside Sydney. Tebbutt initially thought that it was a nebula, because the object did not move. He was able to re-observe the 'nebula' on 17 May, and could not determine any motion. It was only on 21 May that any change of position became evident. This 'strange' behaviour was caused by the comet's path, which was more-or-less directly toward the Earth.

As early as the end of May, the comet was visible to the naked eye from Australia. At the beginning of June the magnitude was already 2, and a tail 3° long was visible. The brightness subsequently increased significantly. On 8 June the comet was already so bright that it remained visible 40 minutes after sunrise. The tail measured 18°, and by 20 June was 40° long. This was joined by a second, 5° tail.

When the comet became visible in Europe on 30 June, Tebbutt's discovery report had not yet arrived. By this date, the head of the comet

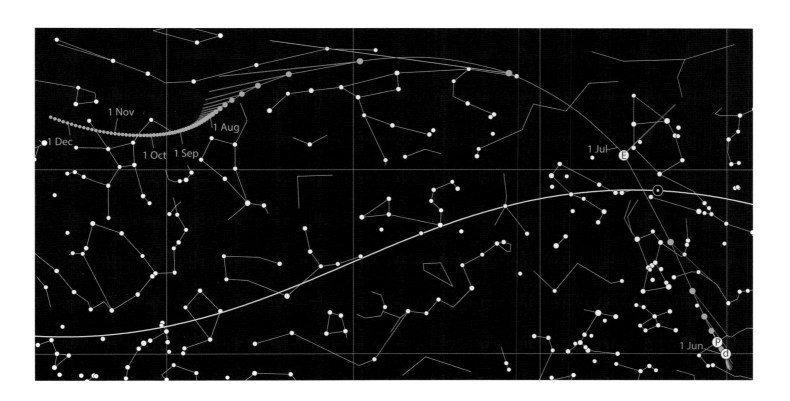

was visible above the horizon both in the morning and in the evening. The tail stretched across half the sky from Auriga to Aquila. The German astronomer, Julius Schmidt, the director of the observatory at Athens, described the view: "The dull yellowish-red head, the size of the Moon, was very close to the horizon; the tail, curving downwards, shining widely, and immediately compared by all the observers present to the distant glow of a considerable fire, initially pointed towards the Pole Star, and could, with ease, be distinguished over a distance of 120 degrees." Schmidt compared the magnitude of the head to that of Jupiter, and spoke of a 'shining tail that cast shadows'. The magnitude must therefore have been greater than –2.

The nucleus of the comet measured about 3' in diameter. In small telescopes there were already up to six nebulous shells (envelopes). In the extremely wide tail, fanning out from the head over an angle of 80°, numerous dark rays could be seen. Its colour, however, was white, but greenish tints were also reported. When the head was below the horizon, a few observers mistook it for an aurora. East of the main tail, a second tail was visible, about 30° long.

On 1 July the magnitude amounted to about –2. The tail was seen to be at least 70° long, but was, however, smaller than the day before. On 2 July the magnitude had decreased to 1. But the length of the tail was now reported to be 100° to 120°. On 3 July, the length observed was similar.

In the following days, decreases in brightness and tail-length were observed, but until 7 July the appearance of the comet was extremely striking. With a magnitude of 2 it was still impressive, and the tail fanned out over 50° of the night sky. Through a telescope the nucleus showed distinct jets and envelopes.

By the middle of July the magnitude had declined still further, and amounted to 3. The length of the tail shrank to 12°. In August, the tail still remained clearly visible, but had shortened to 2.5°. The magnitude of the comet fell to 5.

Telescopically, Comet Tebbutt was visible until well into 1862. The last observation from the Pulkova Observatory in St Petersburg dates from 30 April, when the comet had a magnitude of only 14.

Background and public reaction

Extreme lengths of the tail and great brightness do not occur simultaneously in most comets. Because of its orbital geometry, which brought it close to the Earth, while the comet did not pass particularly close to the Sun, Comet Tebbutt was an exception. John Herschel stated that Comet Tebbutt "exceeded all other comets that I have seen in brightness, even those of 1811 and 1858".

Unlike the case with Comet Halley 50 years later, the Earth's passage through the tail of the comet did not cause any mass panic. In England and Australia, luminous effects in the sky similar to aurorae were noticed when this occurred. These sightings are, however, nowadays regarded as uncertain, because the observers were biased by their expectations to see something during their passage through the tail.

Schmidt observed periodic fluctuations in the length of the tail with a period of 25.6 days and in the diameter of the coma with a period of 25.4 days. Pulsations of the nucleus were also reported, but these were probably only caused by scintillation in the atmosphere. What were real, however, were observations by Fyodor Bredikhin of rotating jets from the cometary nucleus on 30 June. These showed for the first time that these bodies rotated.

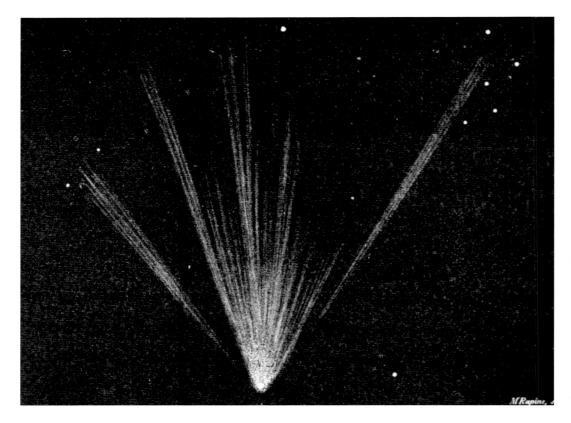

This sketch of 30 Jun. 1861 clearly shows the extent of the tail. The constellation of Cassiopeia may be seen top right, below right is the star Capella, and at the top edge of the picture is the Pole Star.
F. Chambers

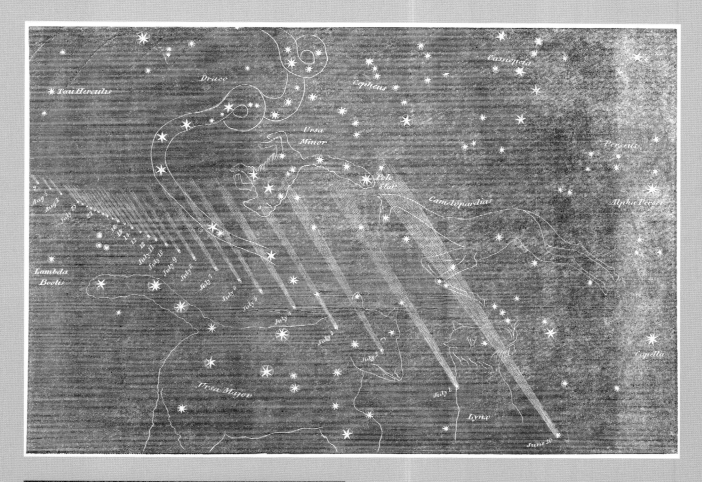

These chart sketches show the apparent path of the comet between the 30 Jun. and 9 Aug. 1861 (*above*). The detailed view of the head (*below left*) is from 30 Jun. 1861, and the view (*below right*) is of unknown date.

✑ Great September Comet of 1882

I ts extreme values remain unsurpassed to this day. C/1882 R1 was, with its estimated magnitude of –17, the brightest comet of modern times. It was so bright that it could be seen without difficulty directly next to the Sun in the daytime sky. Its brilliance was caused by its extreme proximity to the Sun, as it approached to a distance of less than one million kilometres. Subsequently, it developed an impressive, bright tail, which spread magnificently across the southern sky.

▲ The Great September Comet on 6 Oct. 1882. *L. Weinek*

Data

Number:	18
Designation:	C/1882 R1
Old designation:	1882 II, 1882b
Discovery date:	1 Sep 1882
Discoverer:	Unknown
Perihelion date:	17 Sep 1882
Perihelion distance:	0.0078 AU
Closest Earth approach:	17 Sep 1882
Minimum Earth distance:	0.9773 AU
Maximum magnitude:	−17
Maximum tail length:	20°
Longitude of perihelion:	69.6°
Longitude of ascending node:	347.7°
Orbital inclination:	142.0°
Eccentricity:	0.99989400

Orbit and visibility

With a distance from the Sun of less than one million kilometres, the Great September Comet of 1882 was a perfect example of a sungrazer. The extremely close approach to the Sun was not only responsible for its great brightness, but also for the impressive development of the tail. Because of the small perihelion distance, the closest approach to Earth nearly came at the same time as perihelion.

At the time of the first sighting at the beginning of September 1882, C/1882 R1 was south of the Head of Hydra. As such it was visible in the morning sky, where the position was somewhat better for observers in the Earth's southern hemisphere than for those of the northern hemisphere.

From Hydra, in the first two weeks of September, the comet moved directly towards the Sun, which was between Leo and Virgo. However, as it neared the Sun, the visibility conditions significantly worsened.

On 17 September the double conjunction with the Sun took place. To begin with, the comet passed in front of (transited) the Sun over a period of one hour, seventeen minutes. The comet then moved away from the Sun, but turned back at a maximum distance of only 27', and approached the Sun again. The second encounter then occurred, in which the comet was eclipsed by

the Sun for just about two hours. This unusual transit and eclipse took place one after the other on 17 September with just a few hours between them.

After its dance around the Sun at perihelion, the comet moved southwest, which introduced its second period of morning visibility. The more southerly path than before perihelion caused significantly poorer visibility for the northern hemisphere, and only in October could the comet be seen again ahead of the Sun. At southern locations the visibility in the morning twilight was, in contrast, significantly better.

On 1 October, the comet had entered the constellation of Sextans, and on 11 October, Hydra. It continued in a south-westerly direction, and moved into Pyxis in December 1882. In doing so, C/1882 R1 could now be observed throughout the night, however, it was only in southern Europe that it was high enough above the horizon.

In January 1883, the comet passed through the 'legs' of Canis Major, deep in the winter sky. While its path turned north again in the direction of Orion, and it could be seen again from more northern regions, the distance also increased considerably. In February and March, C/1882 R1 became an evening object, where, depending on latitude, it remained visible into June 1883.

Discovery and observations

When the first observations of the Great September Comet took place on 1 September, from central and southern Africa, C/1882 R1 was already a gloriously bright object. An unnamed Argentinian observer described it on 6 September as 'as bright as Venus, with a bright tail'. It can now no longer be established who first saw the comet. Based on its

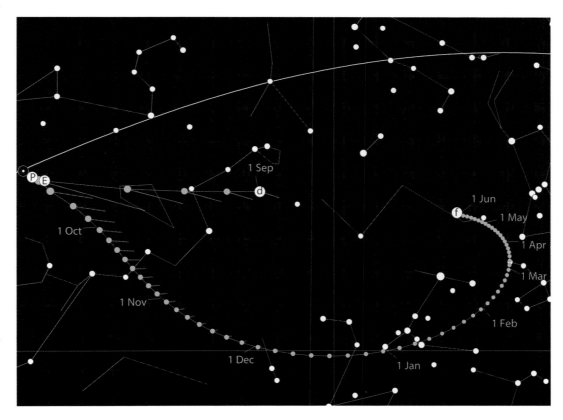

brightness, it was probably the case that many individual people saw the comet at about the same time in the morning twilight.

A week after the first sightings, the South-African astronomer William Finlay first saw the comet. He estimated the magnitude of the head at 3, but the overall brightness must have been considerably greater. At this period a tail of about 3° in length was already visible.

By the 12 September the comet had become so bright that it could be followed until sunrise. On the following day its brightness was compared with Jupiter, so it was already at a significant negative magnitude. The tail was then 12° long. In the following days, despite its decreasing distance from the Sun, it ever more easy to see the comet.

On 17 September at perihelion the comet was finally visible in the daytime sky. As such it appeared right next to the Sun 'so bright, that a glance at the Sun showed the comet without any need for a search'. The comet was much brighter than Venus, and could even be seen through thin cloud in the daytime sky. Its rapid motion past the Sun was noticed even by laymen.

At this phase C/1882 R1 was probably the brightest comet in history. It was certainly the brightest of modern times. With its estimated magnitude of between –15 and –17 it was even visible through a solar filter. Finlay described the appearance: "The silvery light of the comet formed an impressive contrast with the reddish-yellow light from the Sun." William Elkin observed the comet until it disappeared at the solar limb and started to transit the Sun. He only lost the comet right at the ceaselessly fluctuating limb of the Sun.

When in front of the Sun the comet was not visible. After the eclipse that followed the transit, the comet was again visible with the naked eye on 18 September near the Sun. At the time of eclipse it displayed a tail 0.5° long. At this period it was also perceived by a wide section of the general public. The magnitude amounted to about –10.

In the following days the comet again emerged into the morning sky and displayed the most spectacular phase of its visibility. Until 20 September, the comet remained visible in the daytime sky. In the meantime, the tail had lengthened to 5°, and on 22 September was even seen at 12°. It now became longer day by day, and eventually reached 20°. Its width was, however, only 1° and because of its slight curvature it gained an elegant aspect. It remained visible, maintaining this shape, until the end of October 1882. The comet's magnitude at the change from October to November was still an impressive 0.5.

On 27 September, Edward Emerson Barnard observed the consequences of the close passage past the Sun. Through a 5-inch telescope and with a magnification of 78×, the nucleus of the comet appeared elongated. On 2 October it was already egg-shaped, and on the following day, it divided. On 5 October, three fragments of the nucleus were visible. In subsequent days, additional portions of the nucleus became visible, which by 10 October had become aligned into a straight line with an apparent length of 22°. On 15 October five, and later six, nuclei were seen, which offered a changing appearance from night to night, and which appeared like a 'string of pearls'.

In the middle of October, C/1882 was still at magnitude 1.0. Telescopically, five to six nebular shells, known as envelopes, were visible around the nucleus. The innermost of these began in the elongated, multi-cored centre of the comet. From the end of September, a faint luminosity could be detected around the comet's tail, and which was not curved like the main tail but was straight. It appeared brighter at its edges, dark in the centre, and was most striking in October. An anti-tail appeared to develop out of this luminosity, but both phenomena had disappeared by 17 October.

On 5 October, a patch of nebulosity, detached from the main portion of the comet, was seen for the first time. It could only be detected under very dark skies and was 1.5° away from the main body, but rapidly receded. On 10 October it appeared like a faint second comet. In the following days, Barnard was able to make out a total of six further objects, like individual small comets. On 14 October they were up to 6° away from the main comet.

In November, the tail still measured 20° long. Its specific shape, consisting of two unequal arms, resembling the Greek letter γ (gamma), persisted until the middle of December 1882. At the turn of the year, the length of the tail was still more than 10°. At that period, several nuclei were still visible through a telescope.

In February 1883, the comet's magnitude was between approximately 5 and 6. A tail 5° long was still visible to the naked eye. The comet could be seen without any optical aid until the beginning of March. The last observation of 1 June came from the Argentinian astronomer Juan Thome. Benjamin Gould noted that the comet "could no longer be observed, not because of a lack of brightness, but because of its low altitude in the twilight".

Background and public reaction

The magnitude of the Great September Comet of 1882, of between –15 and –17, has never been exceeded by any comet up to the present day. C/1882 R1 is considered to be one of the brightest comets in history. It is the prototype for a sungrazer and, like Comet Ikeya-Seki in 1965, is included in the Kreutz Group, which must have formed from a single parent object that fragmented around the year 371 BC. One piece of it may be identical with the Comet of 1106.

In 1880 and 1887 two more sungrazers appeared, which attained even smaller perihelion distances than that of the Great September Comet. Both were similarly placed in the Kreutz Group, but in another sub-group, however, from C/1882 R1. They went down in history as 'Great Southern Comets', because, during their periods of best visibility, they were in the southern sky. However, neither came anywhere near the brilliance of the Great Comet of 1882. Despite its extremely close approach to the Sun and the fragmentation of its nucleus, the Great September Comet survived perihelion and remained a bright object for a long time. The orbital periods of the fragments were calculated at 671 to 955 years, whereas the main comet should return in 773 years, if it has not disintegrated in the meantime.

C/1882 R1 is the first comet for which successful photographs have survived. Among these are the famous images that David Gill succeeded in obtaining from the Cape of Good Hope. Between 20

October and 15 November, he photographed the comet with a 60-mm portrait lens, mounted piggyback on a telescope. The exposure times ranged from half-an-hour to over two hours. Gill was so fascinated by the many stars shown that, using this arrangement, he began the *Cape Photographic Durchmusterung*, the first photographic star catalogue.

On 18 September, L. Thollon and A. Guoy, during a daylight observation from Nice, were able to detect the Doppler effect in the double line of sodium, which had, shortly before, been identified for the first time in comets. The shift in the spectral lines enabled them to derive a velocity of 61 to 76 kilometres per second away from the Earth.

View of the Great September Comet at the time of its greatest brightness. *Anonymous*

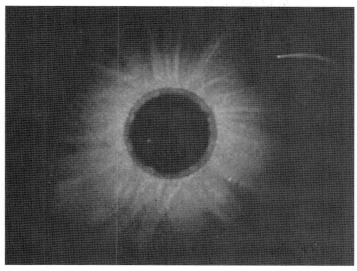

Four months before the Great September Comet, another comet caused a furore. At the observation of the solar eclipse of 17 May 1882 in Egypt, X/1882 K1 became visible during the phase of totality. Shortly afterwards it crashed into the Sun. *A. Schuster (above), anonymous (top)*

This photograph of the comet on 14 November 1882 is among the first successful comet photographs ever taken. *David Gill*

◄ The Great September Comet during the phase of its maximum brightness on 13 September 1882 above the pyramids in Egypt. *Garnet Wolseley*

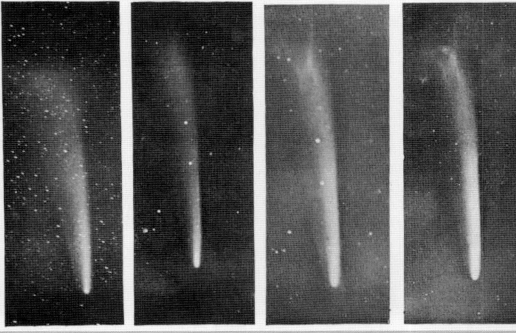

◄ A series of photographs taken at the Royal Observatory at the Cape of Good Hope. *David Gill*

◄ The head of the comet according to a contemporary drawing. *Anonymous*

☄ Great January Comet of 1910

In January 1910, the whole world awaited the widely announced return of Halley's Comet. But suddenly a new comet burst onto the celestial scene, which stole the show from the famous example. The Great January Comet of 1910 was both brighter and better to see than Halley, which arrived three months later. Many people, who later thought they remembered the impressive sight of Comet Halley had, in reality, seen the Great January Comet.

One of the few photographs obtained of the Great January Comet of 1910. *Unknown photographer*

Orbit and visibility

The Great January Comet of 1910 was discovered five days before its perihelion passage. At that time it was in the constellation of Sagittarius, 12° south of the Sun, and was thus invisible to observers in the northern hemisphere. Observers located in the south could, however, see it in the morning sky.

On 18 January, the comet passed north of the Sun at a distance of just 2.8°. In the following days it thus passed into the evening sky for the northern hemisphere. In doing so, it moved from Capricornus into Aquarius. Somewhat farther to the east in Aquarius was the bright planet Venus, as the 'Evening Star'. The comet moved rapidly north, so that visibility conditions significantly improved in European skies. As January changed into February 1910, it crossed the border of the con-

Data

Data	
Number:	19
Designation:	C/1910 A1
Old designation:	1910 I, 1910a
Discovery date:	12 Jan 1910
Discoverer:	Unknown
Perihelion date:	17 Jan 1910
Perihelion distance:	0.1290 AU
Closest Earth approach:	18 Jan 1910
Minimum Earth distance:	0.83 AU
Maximum magnitude:	−5
Maximum tail length:	50°
Longitude of perihelion:	320.9°
Longitude of ascending node:	90.0°
Orbital inclination:	138.8°
Eccentricity:	0.99999500

stellation of Pegasus. The bright Moon interfered with observations at the end of January. In February and March, the comet moved west, past the Great Square of Pegasus, and then northwest into the Milky Way, where all trace of it was lost in July 1910.

Nowadays, the Great January Comet of 1910 is regarded as being a 'fresh' object from the Oort Cloud on its first visit to the inner Solar System. C/1910 A1 has an extremely long, extended orbit. Its return is expected to be in about four million years.

Discovery and observations

Mine-workers in South Africa were the first to discover the comet, already visible to the naked eye in the morning sky, at their change of shift. South-African railway workers also saw the comet. They took it to be Halley's Comet, which, however, at that time was in Aries and was considerably fainter, because it would only reach perihelion in May. As a result, it was some days before professional astronomers explained the confusion and the object was recognized as a new comet.

On 17 January, the comet was so bright that during its passage past the Sun it could be seen with the naked eye as a 'snow-white' object. The magnitude was compared with Venus at its brightest, and must therefore have reached −5. On 18 January, the observatories at Vienna, Rome, Algiers and Milan reported daylight sightings of the comet.

In the following days the best visibility came in the evening sky. From 19 January the comet was visible as a very bright object. The overall magnitude that day amounted to about −2.5. The extremely bright dust tail had a slightly curved shape and a yellowish or reddish tint. Its length was given on 20 January as 7°, on 22 January 25°, and on 30 January as much as 50°. On 28 January, E. Silbernagel from the Munich Observatory reported: "nucleus about magnitude 1.5. Tail not particularly wide, but very long, however: at about half-8 [7:30] could be followed for about 50°. Towards the end it was bent towards the south and widened into a mist".

The pronounced markings on the tail by striae were particularly remarkable. A fainter gas tail formed after 26 January. In February, the beginnings of an anti-tail could be detected. But the comet's magnificence decreased just as rapidly as it had increased: at the end of the month the brightness abruptly declined, and after 5 February the comet could no longer be seen with the naked eye. By the middle of February, the magnitude was 6, and at the beginning of March just 8. The last sighting of the comet by Max Wolf from 15 July 1910 was reported at a magnitude of 16.5. By that time, C/1910 A1 was already at a distance of 3.4 astronomical units from the Sun.

Background and public reaction

At a time when Comet Halley had already been in the sky for several months and everyone was preparing for the comet's close approach to Earth in May 1910, the Great January Comet appeared as a completely unexpected harbinger of Comet Halley. It caused an exceptional increase in awareness, but also increased expectations for Comet Halley. In the end, however, the famous comet failed to reach the brightness and beauty of the Great January Comet. Many contemporaries, who saw a bright comet in 1910, confused it in their memories with Comet Halley, which first became an object visible to the naked eye in April 1910, when the Great January Comet had already vanished from unaided sight.

The Great January Comet of 1910 is one of the comets that was visible for the shortest period with the naked eye. After perihelion, this lasted for a period of two weeks, in which the comet was a bright, impressive object for just one week.

C/1910 A1 was a very dust-rich comet, which promoted the formation of a beautiful tail. Spectroscopically, clear sodium emission was detected. The gas emissions, typical of comets, only began when the sodium radiation began to subside.

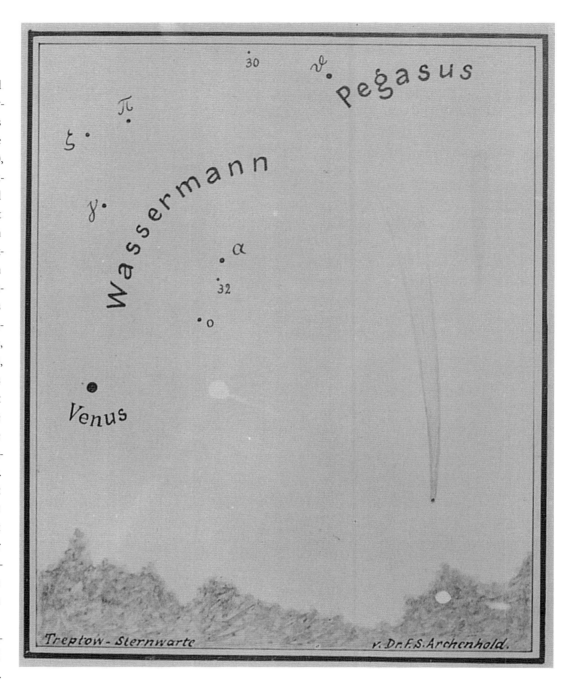

▲ View of the comet over the Berlin night sky at the end of January 1910. *F. S. Archenhold*

▼ Detailed sketches of the head of the comet through the great telescope of the Treptow Observatory in Berlin. *F. S. Archenhold*

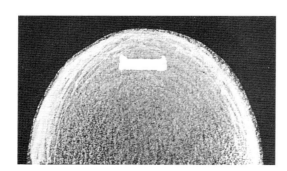

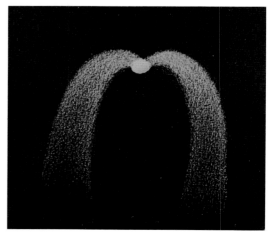

Comet Halley 1910

H istorically, Halley's Comet had often caused a sensation, but at no return had the excitement been so great as in 1910. At a time of enthusiasm for technology and astronomy, when many people accepted extraterrestrial life existed on Mars, and astrophysics was about to enlarge the picture and extent of the cosmos, the announcement that the Earth would pass through the tail of Comet Halley and its poisonous gases, precipitated an unprecedented reaction. But the end of the world was cancelled and afterwards many were left disappointed at not having seen the comet.

Halley's Comet on 21 April 1910, photographed at the Boyden Station near Arequipa in Peru.

Data

Number:	20
Designation:	1P/Halley
Old designation:	1910 II
Discovery date:	11 Sep 1909
Discoverer:	Max Wolf
Perihelion date:	20 Apr 1910
Perihelion distance:	0.5872 AU
Closest Earth approach:	20 May 1910
Minimum Earth distance:	0.15 AU
Maximum magnitude:	−3
Maximum tail length:	240°
Longitude of perihelion:	111.7°
Longitude of ascending node:	57.8°
Orbital inclination:	162.2°
Eccentricity:	0.9672968

Orbit and visibility

The recovery of Halley's Comet occurred in the constellation of Ge-mini, not far from the star γ Geminorum. After September 1909, the comet subsequently moved through the constellations of Orion and Taurus, which it reached in December. In doing so, during this early phase the comet was ideally placed, opposite the Sun in the dark night sky.

In January 1910, the comet had entered the evening sky in the constellation of Aries. Until the 12 March it was visible there from northern latitudes. On 25 March it was overtaken by the Sun in Pisces, and was thus unobservable worldwide, being behind the daytime star.

At the beginning of April, still in the constellation of Pisces, the second phase of visibility began in the morning sky. On 3 May, the comet appeared with Venus and the crescent Moon in the morning twilight. On 7 May it reached its maximum separation from the Sun during this phase of visibility with an elongation of 41°. Until 16 May there were passable observing conditions in the morning sky.

After that Comet Halley rapidly approached its second conjunction with the Sun. This time it was ahead of the latter and on 19 May around 05:00 Universal Time it passed in front of the Sun's disk. In the evening and night of 19–20 May, the sight geometry was such that the tail was visible in both the evening and morning, because that day the Earth almost passed through the tail. The dust tail was then missed by about 300 000 kilometres, while the Earth actually passed through the gas tail.

During the remainder of May, the comet moved parallel to the ecliptic through Taurus, Gemini and Canis Minor. In June it was south of Leo, and in July reached Sextans, where its evening visibility ended. In autumn 1910 it was visible in the morning sky in the constellations of Crater and Corvus. The last sighting showed it in the constellation of Sextans.

Discovery and observations

Ever since Palitzsch's master stroke in 1758, the recovery of Halley's Comet was regarded as a matter of prestige. It was obvious that a twentieth-century technique, photography, which had not been discovered 75 years earlier, would be brought into play. The race was won by Max Wolf, one of the leading astrophotographers of the day. Using the 28-inch Waltz telescope at the Heidelberg Observatory on the Königstuhl, on 11 September 1909, he exposed a photographic plate for an hour. The comet appeared on it as an object of magnitude 16.

Later, the comet was also found on exposures that had been made earlier by the Helwan Observatory in Egypt on 24 August, and Greenwich Observatory on 9 September. The first visual sighting was obtained by the American astronomer Sherburne Wesley Burnham with the 40-inch refractor at the Yerkes Observatory in the USA on 15 September. At the time the comet was still faint, at magnitude 15.5.

The magnitude rose to 12 by the middle of November and to 10 a month later. In February 1910 it was 7. Max Wolf was the first to see the comet with the naked eye on 11 February, but no other observer apparently succeeded in doing so before the first passage past the Sun.

After solar conjunction at the beginning of April, Halley's Comet was initially visible only telescopically. It was towards the middle of the month before it could be detected with the naked eye. At first, it appeared as a 'reddish star', probably because of its location very close to the horizon. However, the magnitude rose over the course of the month to 2. The tail could be seen to a length of 2.5°. At the beginning of May this value increased continuously. Edward Emerson Barnard estimated the length of the tail, under dark skies, as 18° on 17 May, 53° on 14 May, and an unbelievable 120° on 18 May, immediately before passage past the Sun. From the Sonnwendstein (altitude 1523 metres) in Lower Austria, a length of 32° was determined on 12 May, 83° on 17 May and 140° on 19 May. At this period the comet was, overall, at its most conspicuous. It was eye-catching alongside Venus, with the tail stretching upwards like a finger.

On 11–12 May, the first magnitude outburst occurred, which took the total magnitude to 1. This value for the comet's magnitude also persisted when it moved on towards the spectacular solar conjunction and the subsequent passage of the Earth's orbital plane.

The transit in front of the Sun itself, could obviously not, despite many assertions to the contrary, be observed. The comet was indeed seen in the daytime sky shortly before and shortly after, with the assistance of small optics, if the Sun itself was hidden. As such, the magnitude must have been much brighter than the value of 0 that is generally quoted. Estimates speak of up to −6, which is undoubtedly significantly too great, and the value is more likely to have been about −3. This greatly increased magnitude in the days around solar passage may be explained by forward scattering of sunlight by the cometary particles. This effect was fostered by the Earth's passage through the tail in subsequent days. Nothing of this was, however, noticed by the general public.

The threatened passage of the Earth through the tail had no consequences, although Max Wolf believed that he had noticed 'marked cloudiness' by cirrus and reported a particularly intense twilight – and other observers gave similar reports, but none of them can be shown to be linked to the comet.

The location of the Earth on 20 May had another consequence, however. The comet's tail could be seen in both the morning and evening skies. According to Eugène Antoniadi, the total length of the comet's tail was then 240°. There are even a few sightings that suggest a complete, full span of 360°. This is feasible, on the basis of the scattering of light by the tail particles in the antisolar direction, when the Earth was in the tail itself. Such observations would, however, only be possible under a very dark sky.

In the days following when the comet was again in the evening sky, the tail-length decreased from 140° on 20 May to 65° on 26 May. In doing so, the tail rapidly became fainter, and could be distinguished only from dark sites. On 23 May there was a lunar eclipse over America, during which the comet was seen.

The value for the overall magnitude also decreased rapidly from 2 to about 6 on 1 July. After 30 May there was a further outburst, which was linked to a split in the nucleus, which the Fabra Observatory in Barcelona was able to photograph on 2 June. It was probably also because of this event that astonishingly large tail-lengths remained visible for such a long time. At the end of May, 50° was reported, and 15° even as late as 9 June.

Without optical aids, the comet could be seen until June 1910. The last visual sightings date from 1911. The last photograph was successfully obtained by Heber Curtis on 16 June 1911, at a magnitude of 18.

Background and public reaction

The excitement that accompanied the turn of the century still reverberated at the time of Comet Halley's return in 1910. It came at the time of Mars hysteria, inflamed by the martian 'canals' of Giovanni Schiaparelli of 1877, and re-ignited by Percival Lowell at the beginning of

the twentieth century. The appearance of a comet had a lasting effect for decades, and was the trigger for numerous works of science fiction.

Halley's Comet reflected the anxieties about the past and also about the future that were linked to comets. When the British king, Edward VII, died in May 1910, the event was quickly linked to the comet. The simultaneous warning of a meteor storm, on the other hand, was a reflection of the scientific results of the nineteenth century, which had established a link between comets and meteors.

The Earth's passage through the tail of Halley's Comet engendered major fears. After cyanogen had been detected in Comet Morehouse in 1908, a lot of people were afraid of the deadly hydrocyanic acid (prussic acid) that might enter the Earth's atmosphere. After cyanogen was observed spectroscopically in Comet Halley at the end of 1909, this unleashed a wave of hysteria, particularly in France. Sales of gas masks and bottles of oxygen flourished, and at the peak of the panic on 18 May there were even isolated cases of suicide. There was disquiet elsewhere as well: mine-workers in Pennsylvania refused to enter the mines; and in other parts of the USA 'comet pills' were best-sellers. Yet poisonous hydrogen cyanide is present in comets in extremely low amounts, and any influence of the tail on the Earth's atmosphere may be essentially neglected – at any rate, effects cannot be established. Many enlightened contemporaries made fun of the comet fears. In Germany, for instance, postcards with a comet theme and captions such as 'The Comet is Coming, the End is Nigh' sold like hot cakes.

Overall, the general public was disappointed in the sight of Comet Halley. The Great January Comet, with its impressive appearance had stolen the show at the beginning of 1910. Furthermore, bad weather prevented observations in Europe. At the 'End of the World' on 18 May the comet was, in any case, not easily visible, because it was very close to the Sun. In Paris there were boisterous celebrations. Numerous curious onlookers climbed the Eiffel Tower and onto house roofs, but it was largely cloudy. A few unwary comet fans fell from the roofs and were killed.

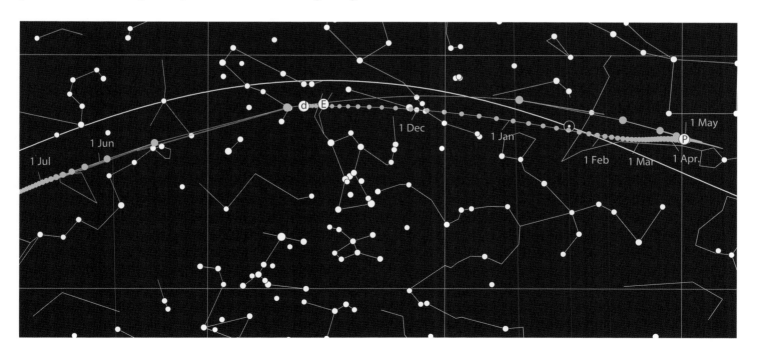

In Cologne there were festivities like those at carnival time. In Vienna there was a 'New-Year' mood, and people gathered at many high points. Practically no reaction was reported from London, although in typical style, it was pouring in rain. At Cape Town people sought shelter, while in New York 'comet banquets' and balloon flights took place. But from Chicago, on the contrary, it was reported that homeowners stopped up the keyholes of their doors, in fear of the poisonous gases. All over the place, street astronomers exhibited the comet through their telescopes. Coins, plates and all sorts of merchandise were on offer.

As early as 1907 there were attempts to recover the comet. Halley's return was predicted by Philip Herbert Cowell and Andrew Crommelin with an accuracy of three days, based on the available data from its earlier reappearances. The prediction was astoundingly good with its calculation of the perihelion date as 17 April 1910. For this, Cowell and Crommelin were awarded a prize by the German astronomical society, Astronomische Gesellschaft.

Scientific cometary research was at a significantly higher level at Comet Halley's appearance than with previous comets. New, large telescopes in the USA (at Lick, Yerkes and Mount Wilson) and the new techniques of photography and spectrography enabled new methods to be used and gave more refined results.

The American astronomer Edward Emerson Barnard stated that the comet should be monitored, worldwide, without a break, and suggested observations from around the Pacific in particular. In fact, over 60 observatories on all the continents took part in the international campaign. In particular, a vast number of photographs were collected and could be evaluated together, to investigate the mysterious outburst and detachments of the tail. It was thus that, among other results, Karl Schwarzschild and Erich Kron detected fluorescence as the type of radiation in gas tails.

Two typical caricatures of the commotion about Halley's Comet. In Germany, above all, postcards about the end of the world were popular (*right*), whereas in England, the excitement among amateur and professional astronomers was made fun of (*below*). *A. Schultz (right), W. Heath Robinson (below)*

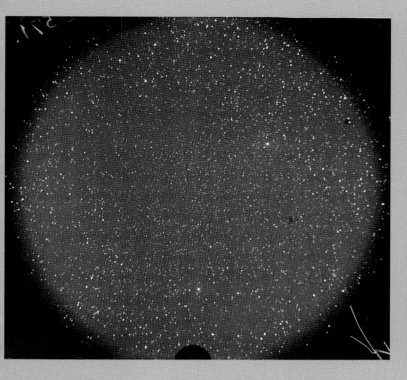

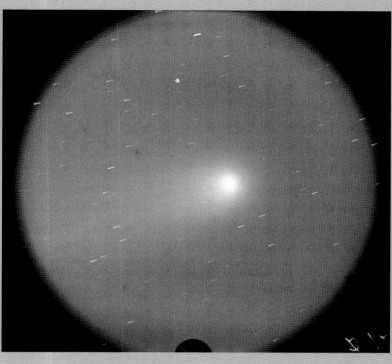

▲ Discovery photograph of 12 September 1909 (*above left*) and detail of the comet's head on 31 May 1910 (*above right*). *Max Wolf*

▼ A sequence showing the development and recession of the tail, photographed from Hawaii. *F. Ellerman*

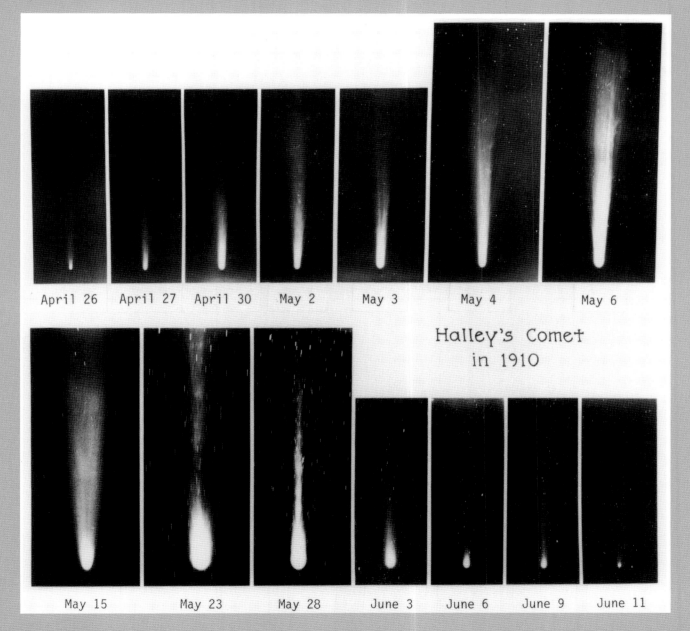

April 26   April 27   April 30   May 2   May 3   May 4   May 6

Halley's Comet in 1910

May 15   May 23   May 28   June 3   June 6   June 9   June 11

Detailed views of the tail on 25 May 1910 (*above left*), 29 May 1910 (*above right*), 11 May 1910 (*below left*) and 30 May 1910 (*below right*). *Unknown photographers (top), H. Knox-Shaw (below)*

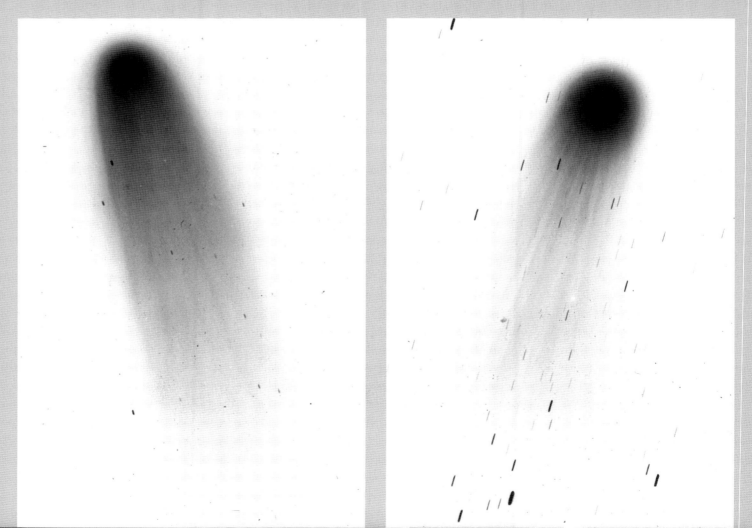

▲ The head of the comet, photographed on 8 May 1910 with the 60-inch reflector at Mount Wilson Observatory.

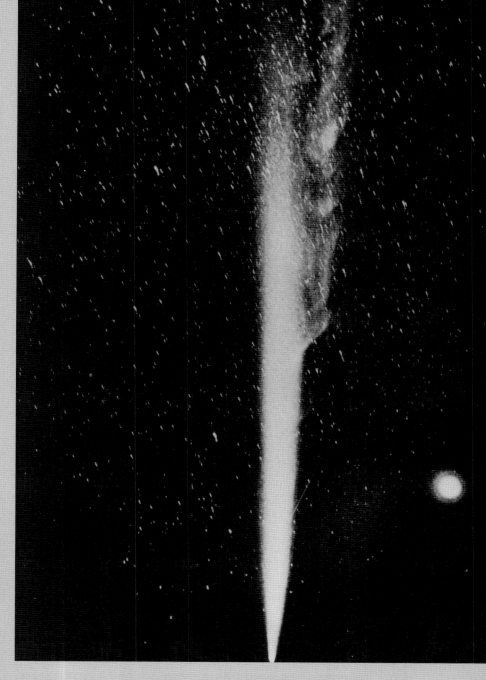

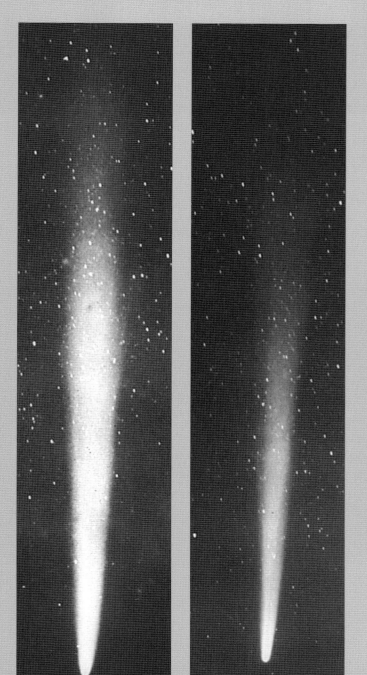

Different views of the tail. The bright object on the right, near the comet in the picture above is the planet Venus.
Unknown photographers

Comet Arend-Roland on 22 April 1957. *Unknown photographer*

Data

Data	
Number:	21
Designation:	C/1956 R1 Arend-Roland
Old designation:	1957 III
Discovery date:	8 Nov 1956
Discoverers:	Sylvain Arend, Georges Roland
Perihelion date:	8 Apr 1957
Perihelion distance:	0.3160 AU
Closest Earth approach:	20 Apr 1957
Minimum Earth distance:	0.5691 AU
Maximum magnitude:	−0.5
Maximum tail length:	30°
Longitude of perihelion:	308.8°
Longitude of ascending node:	215.9°
Orbital inclination:	119.9°
Eccentricity: ·	1.00024440

comet fans in the northern hemisphere had to wait nearly 50 years until Arend-Roland 1956 provided a bright comet to illuminate the sky. Certainly there had been three comets, C/1927 X1 Skjellerup-Maristany, the Great Southern Comet of 1947, and the Eclipse Comet of 1948, all brighter than magnitude 0, but none of them became a striking object worldwide. Skjellerup-Maristany was discovered shortly before perihelion, and the other two only after perihelion. All three were impressive sights only in the southern sky. That Arend-Roland was similarly notable, is for a special reason. Its anti-tail remains to this day a model example.

Orbit and visibility

Discovery took place in the constellation of Triangulum, in the evening sky, when the comet was still very far from the Sun, and could be observed outside twilight. Initially, the comet moved even farther away, until on 4 February 1957 it was at a distance of 1.9 astronomical units. In doing so, it moved south across Pisces into Cetus. This was where its first conjunction with the Sun took place on 20 March, at a minimum separation of 13.7°.

At the beginning of April, a short period of visibility in the southern sky began, before the comet once again approached the Sun – this time moving north. The second solar conjunction came on 16 April, and this time the separation was just 5.2°. A few days later the comet reappeared in the European evening sky. This began the period of best visibility, as the comet moved north from Pisces. Closest approach to Earth occurred not far from the comet's position at discovery.

At the end of April the comet entered the constellation of Perseus, still moving north-westwards. On 15 May it reached its northernmost point, in the constellation of Camelopardalis. The comet was then circumpolar and visible throughout the night for northern Europe. The later path remained at northern declinations and passed through Ursa Major into Draco.

Discovery and observations

The two Belgian professional astronomers Sylvain Arend and Georges Roland were searching for minor planets with the 400-mm telescope at the Royal Belgian Observatory, when something nebulous appeared on a photographic plate, taken with a 50-minute exposure. They first noticed their find in the week following the exposure, during inspection of the plates, and waited for a separate photographic observation before they announced their find. Later, a Japanese photograph of 7 November was found, on which the comet was also visible.

At the time of discovery, C/1956 R1 was faint: about magnitude 10. As early as 27 November, a tail 4' long was identified visually from the Hamburg Observatory at Bergedorf. By the end of December this had stretched to 8', but the brightness had hardly risen.

By the end of January the magnitude had risen to 9. The tail now measured 10' long. A coma, with a diameter of 1', surrounded a faint, stellar nucleus of just magnitude 12. After 4 February, the comet developed into a more prominent phase: its magnitude rose to 7.5 at the end of the month, and the tail elongated to 30'.

In March, Arend-Roland was invisible because of its decreasing distance from the Sun. Renewed observations came after 2 April at a separation of 18° from the Sun. In April the brightness of the comet amounted to about magnitude 2, but because of unfavourable conditions between the two conjunctions with the Sun, the tail was only observed to a length of 5°.

After the second solar conjunction, the comet was again visible from the northern hemisphere after 21 April. The magnitude now reached its maximum value of about 1, and in the last ten days of April the comet was a striking object. Estimates described the tail-length as between 25° and 30°. After 22 April the famous 'spike-like' anti-tail was seen. With a length of 15°, it appeared almost half the length of the true dust tail. It disappeared towards the end of the month. The cometary nucleus had a distinct yellow colour. On 27 April it appeared to be doubled.

At the end of April and beginning of May, the comet underwent a distinct decrease in brightness. The comet faded from magnitude 3 to 7, and the tail shortened from 7° to 1°. In doing so, Arend-Roland became a binocular object.

By the middle of August 1957, the magnitude had faded to 12, and at the end of September 1957, 18 was recorded. The last photograph is from 11 April 1958, taken at a magnitude of 21. On that date the comet, at a distance of 5.5 astronomical units, was already beyond the orbit of Jupiter.

Background and public reaction

Arend-Roland remains today the textbook example of the formation of an anti-tail. No brighter comet has shown this phenomenon in such an impressive manner.

Despite the good evening viewing conditions at the end of March 1957, popular reaction to Comet Arend-Roland was minimal. The low surface brightness of the tail contributed to this, because it only showed to good effect under dark skies.

There was yet another bright comet with a beautiful tail waiting in the wings in 1957. C/1957 P1 was discovered at perihelion by the Czech astronomer Antonín Mrkos, who had already made his name through comet discoveries. The comet had a magnitude of 1, and exhibited a dust tail 13° long and a gas tail 10° long. The particularly complex structure of the gas tail left a lasting impression.

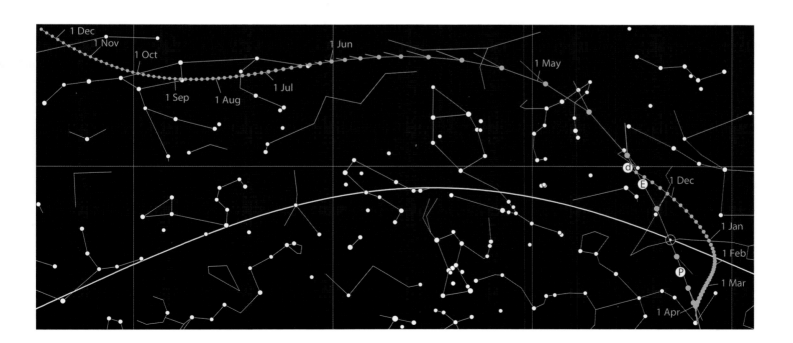

Views of Arend-Roland: 24 April (*above left*) and 28 April 1957 (*above right*). *R. L. Waterfield (top), R. Waland (bottom)*

▶ The comet with a pointed anti-tail on 25 April 1957, when the Earth was passing through the comet's orbital plane. *R. Fogelquist, D. Milon (opposite)*

Comet Ikeya-Seki 1965

Comet Ikeya-Seki on the morning of 2 November 1965. *David Thomas*

Ikeya-Seki was, by any reckoning, a comet of extremes. It approached the Sun to a distance of less than one million kilometres, nearer than any other bright comet of the twentieth century. And because of its extremely close distance to the Sun, it developed into the brightest comet of the century. With a magnitude of –15, it appeared ten times as bright as the Full Moon and could be seen directly alongside the Sun. Its narrow, corkscrew tail was, however, best seen in the southern sky, so that the general population in the northern industrialized countries largely missed this swift comet.

Orbit and visibility

Ikeya-Seki was discovered only one-and-a-half months before perihelion. At that date, the comet was south of the Sun in the constellation of Hydra. At northern latitudes it was only visible under very unfavourable circumstances low on the horizon in the morning sky.

In October 1965 took a path through Crater and Corvus towards Spica, the principal star in Virgo. In doing so, visibility conditions improved significantly, and it was easy to observe in the morning sky.

At perihelion on 21 October, the comet described a loop around Spica and the Sun, which at that date was just a few degrees east of the star. As such, for a few days the comet was only visible in the day-

Data	
Number:	22
Designation:	C/1965 S1 Ikeya-Seki
Old designation:	1965f, 1965 VIII
Discovery date:	18 Sep 1965
Discoverers:	Kaoru Ikeya, Tsutomu Seki
Perihelion date:	21 Oct 1965
Perihelion distance:	0.0078 AU
Closest Earth approach:	17 Oct 1965
Minimum Earth distance:	0.91 AU
Maximum magnitude:	−15
Maximum tail length:	45°
Longitude of perihelion:	69.0°
Longitude of ascending node:	347.0°
Orbital inclination:	141.9°
Eccentricity:	0.99991500

time sky. The apparent minimum separation from the Sun amounted to merely 10'!

The loop around the Sun, in which Ikeya-Seki was always behind our daytime star, catapulted the comet back, in almost the same direction from which it had come previously. So it turned back into the morning sky once more. For northern observers, however, the subsequent path was unfavourable, being even farther south.

On 1 November, Ikeya-Seki was in Corvus. It crossed Hydra in November and moved on to Antlia, where it was finally out of sight of European observers. In January 1966, at the time of the last observations it was in the constellation of Puppis in the southern sky.

Ikeya-Seki is classed as a sungrazer belonging to the Kreutz Group. These comets move on orbits that take them very close to the Sun. With Ikeya-Seki the separation from the surface of the Sun was less than one million kilometres. There is a notable similarity with the path of the Great Comet of 1882, which reached an almost identical perihelion distance.

The orbital period is found to be 880 years. Gary W. Kronk has suggested a possible identity with the September Comet of 1106. The probability of a further return of Ikeya-Seki has undoubtedly decreased following fragmentation of its nucleus in 1965.

Discovery and observations

The Japanese amateur astronomers Kaoru Ikeya and Tsumoto Seki were both active in searching for comets, when they made their greatest discovery on 8 September 1965. Ikeya had already discovered a comet in 1963 with a home-made telescope. His last discovery to date, on 13 November 2010, makes him one of the longest active comet discoverers. Seki is, with six discoveries, even more successful than Ikeya. In addition, he has recovered 28 comets. In 1965 he used a 90-mm telescope with a magnification of just 15×. Both observers discovered the comet within just 15 minutes of one another. In 1967, they repeated a common comet discovery, but this time Ikeya won by a nose.

At its discovery, Comet Ikeya-Seki was already at magnitude 8. Its brightness grew rapidly, just three weeks later the comet could be seen for the first time with the naked eye. At the beginning of October the tail already measured 1.5° long. As it grazed the Sun, the brightness shot up. On 18 October it had already broken the magnitude 0 mark, and the tail was an imposing 10° long. Observers in the south could observe the comet at considerable heights above the horizon, so that even before perihelion there were reports of sightings from the centre of cities such as Sydney.

Then Ikeya-Seki underwent its extremely close passage past the Sun. On 20 October, keen observers were first able to see the comet with the naked eye. They estimated the magnitude that day as about −10. A tail, 1° long, was visible despite the bright sky background. On 21 October, Ikeya-Seki had become significantly brighter. It could be seen clearly about 1.5° from the Sun, when the latter was hidden. A Japanese observatory estimated the magnitude at ten times that of the Full Moon, which

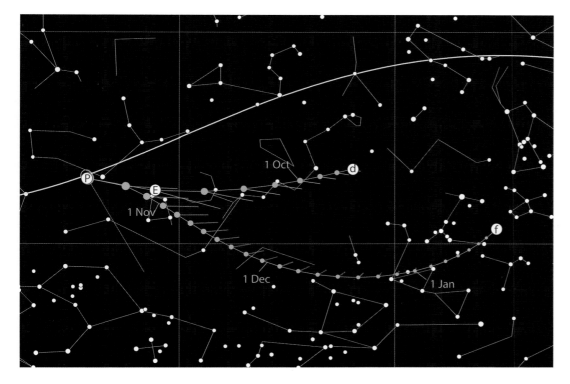

corresponds to about –15. No comet in the twentieth century was brighter. The following day a division in the nucleus was noticed. The comet had broken into three portions, of which only one remained visible subsequently.

The most spectacular sightings of the comet in the bright morning twilight were reported from 23 October. A slightly curved and extremely bright tail, 20° long, emanated from a second-magnitude nucleus. Its slender shape was fascinating. In the following days, as the comet became easier to see under dark skies, the length of the tail doubled to 30–45°. The first half of the tail was nearly as bright as the nucleus. The corkscrew structure recorded on photographs show the so-called 'striae', which arose from ray structures in the tail. Occasionally a faint gas tail could also be detected.

Until the beginning of November, the very bright tail remained unchanging. The experienced comet observer John Bortle reported on 31 October that the comet, despite cloud that blotted out practically all brighter stars, could be seen without interruption. At the same time, however, the magnitude of the comet rapidly dropped, until at the beginning of November it was just 3.

After 4 November a second break-up of the nucleus was observed. Two fragments of the nucleus could now be seen, persisting into January 1966. In December 1965 the tail finally became significantly fainter. The head of the comet, at magnitude 7, could now no longer be seen with the naked eye. The tail now disappeared very rapidly, and by New Year 1965–66 it had already become a difficult telescopic object. The last visual sighting of the comet through a telescope dates from 31 January 1966.

Background and public reaction

Ikeya-Seki was the brightest comet of the twentieth century. Because of its late discovery as well as being exclusively visible in the morning sky, the media were late in finding out about it, and no worldwide enthusiasm was engendered. A hindrance was the relatively short period of maximum brightness. For astronomers in the southern hemisphere, however, Ikeya-Seki remained an event unequalled until the appearance of Comet McNaught in 2007.

In the forecasts for the visibility of Ikeya-Seki, much was made of the Great September Comet of 1882, which came equally close to the Sun. Three years before, in 1962, another comet, Seki-Lines, had already appeared, and passed a short distance away from the Sun. Approaching the Sun to within 0.03 astronomical units, Seki-Lines had reached a magnitude of –7.5 and exhibited a tail 20° long. However, the impression it left did not, by any means, reach that of Ikeya-Seki.

▲ Snapshot on 27 October 1965 from Argentina, where the comet was higher in the morning twilight. *Unknown photographer*

▼ On 21 October 1965 the comet passed the Sun. *F. Moriyama and T. Hirayama*

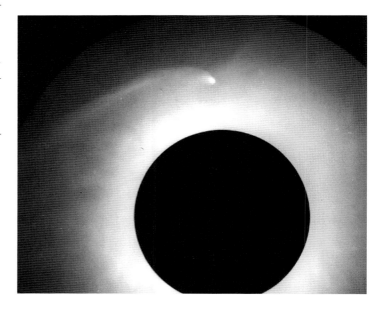

Comet Ikeya-Seki on 29 October 1965. *Unknown photographer*

On 29 October 1965 Comet Ikeya-Seki was at its peak. *John Laborde (left), Roger Lynds (right)*

Comet Ikeya-Seki over the morning horizon on 27 October (*below*) and 29 October (*above*). *Unknown photographers*

☄ Comet Bennett 1970

It was perhaps the most beautiful comet of the twentieth century but, above all, it was one of the least observed. When Comet Bennett was in the morning sky in spring 1970, only a few early risers took any notice of it. Most people slept through it. Three years later, people wanted to make up for the missed opportunity, but Comet Kohoutek in 1973–74 was unable to match the splendour of Comet Bennett.

Comet Bennett on 5 April 1970. *Dennis Milon, Helen and Richard Lines*

Orbit and visibility

C/1969 Y1 was discovered deep in the southern sky near the Small Magellanic Cloud in the constellation of Tucana. Until perihelion on 20 March 1970 it remained an object exclusively for inhabitants of the Earth's southern hemisphere, who were able to observe it in the evening sky.

In February 1970, the comet crossed the constellation of Grus, moving north. Passing west of Fomalhaut, the principal star in Piscis Aus-

trinus, it moved towards the Sun, which it reached on 21 March in the constellation of Aquarius.

After perihelion at the end of March, came the start of visibility for observers in the northern hemisphere. The comet was then visible on the morning horizon in the east.

The phase of best observational possibilities, shortly after closest approach to Earth on 26 March, found the comet in the constellation of Pegasus, west of the conspicuous Great Square of Pegasus. The tail pointed in the direction of the Milky Way.

Data

Number:	23
Designation:	C/1969 Y1 Bennett
Old designation:	1970 II
Discovery date:	28 Dec 1969
Discoverer:	John Bennett
Perihelion date:	20 Mar 1970
Perihelion distance:	0.5376 AU
Closest Earth approach:	26 Mar 1970
Minimum Earth distance:	0.69 AU
Maximum magnitude:	0
Maximum tail length:	25°
Longitude of perihelion:	354.1°
Longitude of ascending node:	224.7°
Orbital inclination:	90.0°
Eccentricity:	0.99619300

Towards the middle and end of April, the bright Moon interfered with observations. At the end of that month, the comet entered Cassiopeia and crossed the autumn Milky Way. It remained there until June, and from September it was to be found in Cepheus. The orbital period of the comet was given as 1680 years. As such, its last perihelion must have occurred in AD 290. No observations have, however, been definitely ascribed to that return.

Discovery and observations

The South African amateur astronomer John Bennett had already spent 333 hours in systematically searching for comets, before he found his first comet on 28 December 1969. Bennett had observed comets since 1960, and began a dedicated search in 1967. Like Messier, he had compiled a list of nebulous deep-sky objects that might be confused with comets, and which might hinder his search. His catalogue of 152 entries, may be considered today as the 'Messier Catalogue of the Southern Sky'.

Bennett saw the magnitude 8.5 comet with his 120-mm richfield refractor, with which he also compiled the catalogue of nebulae. By the end of January 1970, the magnitude of the comet had risen to 7. Up to this point, a tail could only be detected photographically.

In February the magnitude shot up from 6 to 3. By the middle of March, the tail had grown to a

hefty 10°. Then the comet was lost for a few days in the glare from the Sun. At the end of March, Comet Bennett reappeared with its striking, curved tail. Its length amounted to 10–12°. Initially, it was conspicuous for its yellowish or orange colour. Telescopically, in the head of the comet, jets could be detected on the side of the nucleus that was turned towards the Sun, as well as the resulting envelopes. A dark 'nuclear shadow' could also be observed in the tail.

At closest approach to Earth on 26 March, the magnitude of the comet reached its maximum value of about 0. The very bright and highly structured dust tail could be seen to a length of 12°, and individual estimates even put it as long as 25°. A gas tail could be detected from time to time. It 'flickered' in response to changes in the solar wind. Most observations and photographs of the comet were made in the days following, around the end of March and the beginning of April.

In April the magnitude declined to about 5. From May onwards, the comet could no longer be seen with the naked eye, and towards the end of the month the magnitude had shrunk to 9. Telescopically, a tail 2.5° long could still be detected.

In September 1970 the magnitude had dropped below 13, and in January 1971 it amounted to 18.9. The last photograph of the comet on 27 February 1971 showed it at a distance of 4.9 astronomical units, close to the orbit of Jupiter.

Background and public reaction

C/1969 Y1 was the first comet since the appearance of Halley's Comet in 1910 that became evident, worldwide, as a bright, impressive object. Unlike Ikeya-Seki it could be seen in all its glory outside twilight, and was, during the period of best visibility, brighter than Arend-Roland in 1956. Because of its 'classic' curved tail and its position, high above the morning horizon, near the Milky Way, it was

considered to be the most beautiful comet of the twentieth century by many contemporaries.

The general public first took notice of Comet Bennett when it had already passed perihelion and closest approach to Earth. Because, by the end of April, it had already become much fainter, most people had only a short time to observe it. Its location in the early morning sky was another reason that Comet Bennett gained hardly any attention.

The 'missed' Comet Bennett was a contributing factor to the Kohoutek hype three years later, because the general public eagerly accepted the announcement of a new, even brighter, comet, that was supposed to appear in the evening sky. Bennett's display, which greatly exceeded expectations, contributed to the fact that hardly any attention was given to the doubts that were voiced.

▲ Views of the comet on 31 March 1970 (*top*) and 5 April 1970 (*above*). *J. Shuder (top), W. Sorgenfrey (above)*

▲ At the time of these photographs on 11 April 1970 Comet Bennett's maximum brightness was over. *K. Brandl*

Fine rays in the gas tail and a bright, curved dust tail are shown in this photograph of 21 March 1970 from the South African Boyden Observatory.

☄ Comet Kohoutek 1973/74

The famous American comet researcher Fred Whipple once said: "If you must bet, bet on horses – not on a comet." In 1973 hardly anyone took this advice seriously: shortly after its discovery, when it was extolled as being a possible 'comet of the century', media worldwide readily seized on the excitement about Comet Kohoutek and portentously announced a special spectacle to the public. More realistic doubts were unable to stop this machinery, and so expectations and reality diverged inexorably. The great crash came shortly before Christmas 1973. The fact that the spectacle failed to materialize served as a reproach to astronomers for decades.

Comet Kohoutek was much ridiculed as the fiasco of the century, but for science it was an extremely productive object. This photograph from the 48-inch Schmidt Telescope at Mount Palomar shows the comet on 12 January 1974.

Data	
Number:	24
Designation:	C/1973 E1 Kohoutek
Old designation:	1973 XII (1973f)
Discovery date:	18 Mar 1973
Discoverer:	Luboš Kohoutek
Perihelion date:	28 Dec 1973
Perihelion distance:	0.1382 AU
Closest Earth approach:	15 Jan 1974
Minimum Earth distance:	0.81 AU
Maximum magnitude:	−3
Maximum tail length:	25°
Longitude of perihelion:	37.8°
Longitude of ascending node:	258.5°
Orbital inclination:	14.3°
Eccentricity:	1.00000800

Orbit and visibility

The apparent orbit of Comet Kohoutek ran close to, and almost parallel with, the ecliptic. Its visibility was thus divided into two sections, before and after its encounter with the Sun at perihelion on 28
December 1973.

When first sightings of the comet with the naked eye came in November 1973, the comet was in the constellation of Corvus in the morning sky. In December, Comet Kohoutek passed through Libra and
Scorpius as it approached the Sun, which it encountered in Sagittarius. The minimum separation from the Sun on 28 December amounted to just 0.5°.

The comet then moved into the evening sky. In January it passed
through Capricornus into Aquarius, where closest approach to Earth
occurred on 15 January. In February, Comet Kohoutek was in Pisces,
and its distance from Earth increased rapidly. In April, it passed north
of the Hyades.

Before perihelion in 1973, the orbital period of Comet Kohoutek
was calculated as 1.5 million years. After its passage through the inner Solar System, when it was affected by orbital perturbations, the
period was given as just 75 000 years.

Discovery and observations

The discovery of Comet Kohoutek was one of the latest, photographic
comet discoveries from Germany – the very latest occurred on 27 February 1975. It was made by the Czech professional astronomer Luboš
Kohoutek, who emigrated to West Germany after the suppression of
the Prague Spring, and who worked at the Hamburg Observatory at
Bergdorf. Using the 800-mm Schmidt Telescope, which was later moved to Calar Alto in Spain, Kohoutek was searching for the lost Comet Biela. In 1845, this had broken into multiple fragments, and the
last two portions were seen in 1852.

Instead, Kohoutek found a few dozen new minor planets on the
photographic plates. In pursuing these, on 18 March 1973 he came
across a faint, 16th-magnitude nebulosity, which he had captured on
a plate on 7 March. The brightness of the comet increased continuously, although not as rapidly as expected. The first sightings with the naked eye were reported only at the end of November. The last observation before conjunction with the Sun, shortly before the very end of the
year, came on 22 December 1973. At this point, the comet had a magnitude of just 2.8, and so was not a spectacular object to the naked eye.

While on Earth people eagerly awaited the comet's reappearance,
the astronauts of the Skylab mission observed the comet frequently
between 24 and 29 December. They undertook many space-walks and
drew sketches of the comet. There was a conspicuous anti-tail, 3° long.

On 27 December, experienced Japanese amateur astronomers were
able to observe the comet in the immediate vicinity of the Sun, with
binoculars. They estimated the magnitude as about −3.

After perihelion on 28 December, the comet's brightness declined
rapidly. On 1 January 1974, when it reappeared in the evening sky, Comet Kohoutek was at magnitude −1.5, on 4 January it was just 2.5, and
a mere magnitude 4 on 10 January.

The comet displayed a beautiful, structured tail, up to 25° in length.
However, the surface brightness was not particularly high, so that it
showed to best advantage only under dark skies. At the end of Janu-

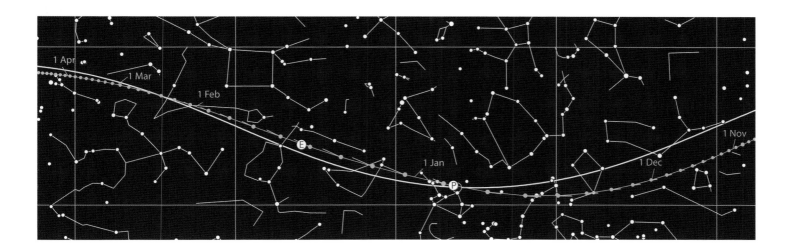

ary the period of naked-eye visibility came to an end. However, with telescopes, Comet Kohoutek could be followed far into 1974.

Background and public reaction

The 'Kohoutek Fiasco' as the media phenomenon was called, has gone down as the greatest astronomical failure of predictions in history, and, to this day, has determined the relationship of the media to astronomical announcements.

The media interest was initially aroused by the exciting stir caused by the early discovery of the comet. It increased when it became obvious that Comet Kohoutek had the potential to become a very bright object – reckoned to be 'a new Star of Bethlehem' by Christmas.

The final trigger for the hype in the USA, however, was when Dave Myers, a NASA expert, spoke to a journalist about the observational programme for Comet Kohoutek, and rather casually remarked that "Comets of this size appear only about once in a hundred years." Overnight, the words became 'Comet of the Century'. Papers announced that Kohoutek could become brighter than Comet Halley of 1910 (which, speaking purely astronomically, was correct), that it would be as bright as the Full Moon in the night sky (which, in historical times, has never been the case with any comet), and would even be seen in the daytime sky (which, likewise, would be correct, with the proviso that only experts, with optical aids, would be able to do so).

The apparent truth of these announcements appeared to be confirmed by the postponement of the Skylab mission so that the astronauts on board could observe the comet during their spectacular EVAs. *Time* magazine featured the comet on its cover, and the luxury liner Queen Elizabeth 2, with hundreds of comet fans and the discoverer on board, put to sea, to be able to catch sight of the comet in the twilight without any light pollution – most of them merely got seasick.

By the beginning of November, it was foreseen that the expectations would not be fulfilled. But the voices of the comet researchers went unheard. In December 1973, sales of telescopes in the USA rose by 200 percent. The crowning success came at Christmas, when President Nixon, in a live broadcast link to the Skylab module, spoke to the astronauts as they observed the comet.

But this project also fell by the wayside like the whole 'Comet of the Century' phenomenon. Shortly before Christmas, the disappointment also finally became obvious to the general public. The hype now changed completely. The 'Wash-out of the Century' and 'Kohouflop' were among the kinder attributes to the comet. American comedians joked that the comet was just a manoeuvre by the government to deflect attention from the Watergate scandal. Another maintained that, this time, a comet had actually predicted the fall of a ruler.

The 'Fiasco of the Century' of Comet Kohoutek has entered popular culture, particularly in the USA. The event was also picked up by the esoteric movement just becoming fashionable that envisaged a new Age of Peace before the final end of the world. The comet also found a positive response among musicians, ranging from Kraftwerk to Pink Floyd. There was even a jazz concert for the comet.

Luboš Kohoutek discovered C/1973 E1 with the large Schmidt Telescope at the Hamburg Observatory. The discovery image shows the comet's motion. *L. Kohoutek*

Against this, Comet Kohoutek was a highly productive object for science. The long preparation time of ten months before perihelion was ideal for researchers. On the basis of the expected brightness, however, a lot of resources were committed, which were not actually employed. Comet Kohoutek was observed from aircraft, high-altitude balloons and from space. While astronauts on board Skylab drew it and made spectroscopic studies, it was also observed by cosmonauts in the Soyuz-13 capsule, and investigated from the interplanetary probe Mariner 10. Comet Kohoutek was the first comet to be studied in the radio region. In doing so, CH_3CHN, HCN, CH, and OH were detected. Evidence for quartz grains (silicates) was found for the first time. Fred Whipple called it "by far the most observed comet in history".

C/1973 E1 was a 'fresh' comet with a shell of light, volatile materials. Because it became active earlier than other comets had done, it was generally assumed that its activity would be higher than its actual level – which accounted for the brightness forecasts of up to –10 magnitudes. After these predictions were found to be incorrect, an additional factor was the fall in brightness after perihelion. This was because the inner, less volatile, nucleus was exposed, and so less material was released than in comparable comets.

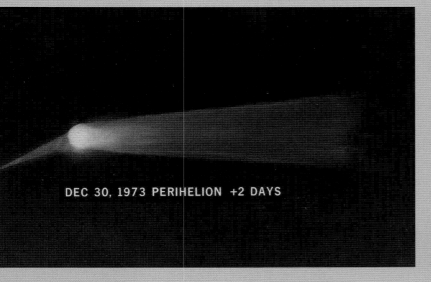

DEC 30, 1973 PERIHELION +2 DAYS

DEC 31, 1973 PERIHELION +3 DAYS

▲ The astronauts on board Skylab drew the appearance of the comet to the naked eye in the days after perihelion on 30–31 Dec. 1974 (*above*). A significant anti-tail is notable.

▼ Photographs of the comet before perihelion from Tautenburg on 22 Nov. 1973 (*left*) and after perihelion from Steward Observatory on 19 Jan. 1974 (*right*).
F. Börngen (left), E. Roemer, L. M. Vaughn (right)

Comet West 1976

M any observers considered this to be the most impressive comet in the previous 40 years. In March 1976, Comet West caught everyone by surprise with its imposing tail. But the Kohoutek fiasco two years earlier, and the visibility in the winter morning sky prevented this comet from improving the status of the cometary scene.

Comet West on the morning of 4 March 1976. *Peter Stättmayer*

Data	
Number:	25
Designation:	C/1975 V1 West
Old designation:	1975n, 1976 VI
Discovery date:	24 Sep 1975
Discoverer:	Richard West
Perihelion date:	25 Feb 1976
Perihelion distance:	0.197 AU
Closest Earth approach:	29 Feb 1976
Minimum Earth distance:	0.79 AU
Maximum magnitude:	−3
Maximum tail length:	30°
Longitude of perihelion:	358.1°
Longitude of ascending node:	118.9°
Orbital inclination:	43.1°
Eccentricity:	0.99997100

Orbit and visibility

At discovery in November 1975, Comet West was in the southern sky in the constellation of Microscopium. It remained in the southern sky and hence unobservable for European observers until February 1976. In the first week of February it passed Fomalhaut, the principal star in Piscis Austrinus, moving north, but was still farther south than the Sun in Aquarius, and thus still remained invisible to Europeans.

Perihelion occurred on 25 February, when the comet's distance from the Sun amounted to less than 30 million kilometres, and was, to a considerable extent, responsible for the following major activity. That day, as seen from Earth, the comet passed the Sun, with a separation of 6° on the sky. In the following days, observers at low latitudes had the opportunity to catch the comet shortly before sunrise and shortly after sunset. With every day that passed, the chances became better, and on the leap-day, 29 February, morning visibility of the comet began, during which it displayed its full splendour.

After 1 March, Comet West moved into the constellation of Pegasus. As it was moving away from the ecliptic, almost at right-angles, towards the northwest, the observational conditions for Central Europe improved rapidly, while the comet could no longer be seen under optimum condi-

tions from the southern hemisphere. The Full Moon only caused interference at the end of March.

In April the comet passed across the constellations of Delphinus and Sagitta and into the summer Milky Way, and in May and June, at an increasing distance, through Ophiuchus. Even before its encounter with the Sun, Comet West was found to have a very long orbital period of 254 000 years. After this return, this was significantly increased to 558 000 years through solar and planetary perturbations to the orbit.

Discovery and observations

The European Southern Observatory (ESO) is a co-operation between European states with the aim of enabling astronomical observations to be carried out in the southern sky. This is, above all, of interest, because of the better climatic conditions. ESO, although headquartered at Garching, near Munich, carries out its work with numerous telescopes in the southern hemisphere, in particular in Chile.

On 24 September 1975, a photographic plate was taken by Guido Pizarro with the 1-m Schmidt Telescope at La Silla, and which was intended for the southern extension of the largest photographic stellar survey at that time, the Palomar Observatory Sky Survey (POSS). The Danish professional astronomer Richard West, working at ESO, examined this plate on 5 November and discovered a small nebulous patch with an incipient tail. Its magnitude amounted to 14–15, and it measured about 2.5' in diameter. The patch was also discovered on plates taken on 10 and 13 August, and proved to be a previously unknown comet.

By the time of its discovery, Comet West had already become significantly brighter. The first visual sightings came as early as November. The brightness of the comet rose further from magnitude 9 in Decem-

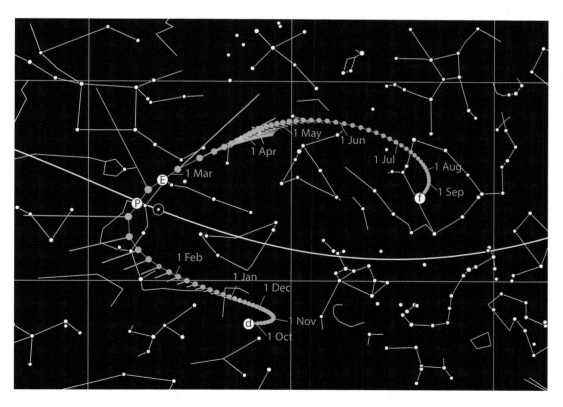

ber to 6 at the beginning of February 1976. At that date a tail 1° long could already be detected, and the comet was visible in binoculars.

In the course of February the brightness increased significantly and had reached magnitude 1 by 20 February. At the end of the month, Comet West was too close to the Sun to be observed before sunrise or after sunset. In fact, many experienced observers succeeded in daytime telescopic observations of the comet on 23 February. Estimates of its magnitude were around –3.

On 29 February observations away from the Sun were again possible. The comet was shining at magnitude –1.5, and rapidly developed a bright tail. From 4 March Comet West could also be seen from Central Europe. The impressive bright tail rose into the dark night sky long before the comet's nucleus, which was reminiscent of reports of the Great Comet of 1744. The complete comet could only be seen in twilight. Light from the tail appeared whitish, and that from the nucleus was golden.

The fascinating, bright and wide tail, which had now been visible for about a week, reached a length of 30° and stretched up into the summer Milky Way. It had a rich structure, with up to 20 synchrones and striae, some of which intersected one another and gave the tail its unmistakeable appearance. A bluish gas tail appeared alongside the dust tail, on the west, and which reached a length of 15°, but which remained significantly fainter.

Located on the morning, western horizon for Central Europe, in the first days of March the tail switched from 'left' to 'right'. On 5 March,

using a 14-inch telescope with a magnification of 250× at the Hoher List Observatory, a second nucleus was seen 3' away from the principal nucleus. On 9 March, two additional nuclei even appeared, and the arrangement of what were now four fragments of the nucleus was reminiscent of the Trapezium in Orion. At the end of March, three nuclei could still be seen. These observations confirmed the comet's fragmentation, which was also confirmed from elsewhere.

On 9 March the magnitude of the comet had declined to 2. On 13 April it was below 6, and in August it had reached 10. The tail was visible for longer than the head. The last visual sighting was on 25 August. On 25 September three cometary nuclei could still be detected, and which were all within a diameter of 1'..

Background and public reaction

After the Kohoutek fiasco of 1974, anxiety about creating a new flop was widespread. While astronomers set their brightness estimates rather too low, the media shrank from the subject. What was possibly the most impressive comet of the twentieth century was, as a result, to a large extent an unnoticed phenomenon, to which its visibility in the early morning sky also undoubtedly contributed.

Comet West was surprising for its significantly greater activity than expected, in particular because there was an additional ejection of dust after perihelion. The origin of this was linked to the fragmentation of the nucleus. The first fracture occurred on 19 February, and a second on 27 February. A third fragment broke away on 6 March. Fragmentation of the comet released enough material to form the simply beautiful tail – its breakup was also responsible for Comet West's greatness.

The break-up of the comet's nucleus in March 1976 (*below*) undoubtedly contributed to its striking appearance in the sky. In the summer of 1976, three nuclei were still visible (*above*).

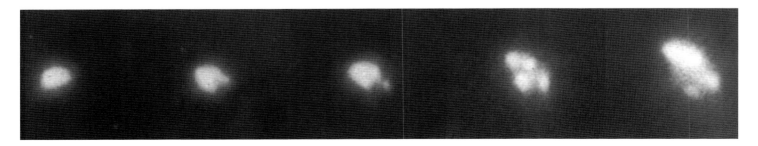

Comet West in the morning twilight on 2 March (*top*) and 3 March 1976 (*centre*). *J. Linder*

▼ On the morning of 4 March 1976, the comet presented a striking appearance over the eastern horizon (*below*). *K. Brandl*

◀ On 9 March 1976, the brightness of the comet had already declined significantly. However, the tail displayed its full splendour outside twilight. *Ronald Royer*

▶ Photographs from the Tautenburg Observatory documented the different views of the tail: 5 March 1976 (*top*), 29 March 1976 (*centre and below*). *R. Ziener, K. Kirsch (top), F. Börngen, K. Kirsch (bottom)*

◀ A fine, bluish gas tail accompanied the imposing dust tail in the days shortly after perihelion. *Jürgen Linder*

Comet Halley 1986

The return of Halley's Comet in 1986 came at a time of great enthusiasm for space. Probes were sent to investigate the Solar System; amateur astronomers could buy cheap telescopes from the Far East; and science fiction was part of popular culture. Unfortunately, Comet Halley's performance would be the most unrewarding in the whole of modern times, thanks to its orbital geometry. So Comet Halley's return prompted a lot of anticipation, which its appearance in the sky was hardly able to satisfy.

Halley's Comet in front of the
Milky Way on 21 March 1985.
Unknown photographer

Data	
Number:	26
Designation:	1P/Halley 1986
Old designation:	1982i
Discovery date:	16 Oct 1982
Discoverers:	David Jewitt, Edward Danielson
Perihelion date:	9 Feb 1986
Perihelion distance:	0.5871 AU
Closest Earth approach:	10 Apr 1986
Minimum Earth distance:	0.42 AU
Maximum magnitude:	2
Maximum tail length:	25°
Longitude of perihelion:	111.9°
Longitude of ascending node:	58.9°
Orbital inclination:	162.2°
Eccentricity:	0.967277

Orbit and visibility

The second appearance of Comet Halley in the twentieth century took place under far less favourable conditions than in 1910. The minimum distance from Earth at 60 million kilometres amounted to twice the value 76 years previously. In addition, the path in 1986 was in the far south, so observers in Europe and North America would only be able to see the comet at the time of its greatest brightness when low on the horizon. Overall, Comet Halley's return of 1986 must be regarded as the most unfavourable of modern times for the inhabitants of Central Europe and North America.

At its extremely early recovery, Halley's Comet was in the winter sky at the borders of the constellations of Orion, Gemini and Monoceros. During the winters of 1982, 1983 and 1984 it described a loop, similar to the loops that outer planets describe around opposition, and which reflect the motion of the Earth around the Sun. In February 1985, Comet Halley was located north of the constellation of Orion.

From autumn 1985, Comet Halley came within reach of amateur astronomers. The comet moved from Orion into Taurus. In October it passed the nebula M1, which Messier had found in 1758 when then searching for Halley's Comet. In November, Comet Halley passed north of the Hyades and south of the Pleiades. At this early stage, far from perihelion, its location in the night sky was ideal, and it could be observed throughout the night. On 27 November its distance from Earth was 0.62 astronomical units. Subsequently, the comet began by moving farther away from the Earth.

In December 1985 the comet again turned west, passing through Aries and into Pisces. At the same time, visibility conditions worsened, and it was only visible in the evening sky. At perihelion on 9 February 1986, in Aquarius, it was hidden by the glare of the Sun, and was thus unobservable.

For observers in the southern hemisphere the comet reappeared in the middle of February in the morning sky, as it was crossing Capricornus in a south-westerly direction. In the period following, from the end of February to the beginning of April, it was reckoned to be at maximum brightness. For observers in Central Europe there was a short, unfavourable window of visibility at the end of March, just above the south-eastern morning horizon, deep in the constellation of Sagittarius. Between 5 and 18 April, and thus during the second close approach to Earth on 11 April, Halley's Comet was invisible for Central-European observers, lying in the southern sky in the constellation of Lupus. The distance to the comet on that day was 0.42 astronomical units.

For observers in the southern hemisphere, however, observational conditions were almost ideal. As it crossed the Tropic of Capricorn, the comet was, at midnight, at the zenith alongside the brightest portion of the Milky Way. At the time of its greatest brightness in the middle of April, Comet Halley crossed the Milky Way from east to west. Subsequently it passed right by the galaxy Centaurus A (NGC 5128).

On 20 April the comet also reappeared in the sky for Central Europeans. In the days following it moved northwards from the constellation of Hydra. The Full Moon on 24 April, however, interfered with observations. This was the period when observers in Central Europe had the greatest chance of seeing the comet.

In May, it turned in the constellation of Sextans and, in doing so, was finally available for all observers, worldwide. However, its distance increased rapidly from day to day.

From Europe only the early and late parts of the event were easily seen. Only at its next return, in 2061, will there be the chance of a better display.

Discovery and observations

As in 1910, early recovery of this famous comet was a matter of prestige. The search was undertaken by all major observatories around the world, but initially, however, without success. American efforts were co-ordinated by the Jet Propulsion Laboratory (JPL) and the California Institute of Technology (Caltech). It was decided to use the 5-m (200-inch) telescope at the Palomar Observatory. There David Jewitt and Edward Danielson had access to a CCD camera, one of the then-revolutionary detector systems.

On the morning of 16 October 1982 they identified the comet as a point-like object only 8' away from the calculated position. At the time, the comet was still 11 astronomical units from the Sun, its magnitude was 24.2 and it appeared without any sign of a coma. It was only two years later, at a distance of 6.2 astronomical units, that the first signs of cometary activity could be established.

Among amateur astronomers there was also a competition to see who would record the comet first. Photographically, this was won by the well-known Japanese comet discoverer Tsutomu Seki on 22 September 1984, when the comet was of magnitude 20.5. Visually, the

American observer Stephen James O'Meara was able to see the comet, using a 24-inch telescope, on 23 January 1985, when Comet Halley was just magnitude 19.6.

In autumn 1985, Halley's Comet became a binocular object at a magnitude of about 7. The first sightings with the naked eye under dark skies followed in November. In December, the comet clearly exhibited a tail, the beginnings of which had been visible earlier. Two components, 4° and 3° in length, were observed.

Comet Halley's behaviour in 1986 was noted for the major activity in the gas tail with numerous disconnection events. The first such even occurred between 9 and 11 January. Two more followed on 9 to 11 March and 11–12 April.

The comet reached perihelion in February, and a magnitude of 2.7 in March. The length of the tail amounted to about 12°, and grew by April to 25°, where the fainter outliers could be detected only under dark skies. A few observers maintained that they had even seen tail-lengths of up to 60°.

At the end of March, because of the viewing geometry, it appeared as if the dust tail made an angle of 90° with the gas tail. During the whole time, the dust tail was broad and diffuse and could not easily be distinguished visually in front of the bright portions of the Milky Way.

At closest approach to Earth in March, Halley's Comet reached its maximum magnitude of 2. That decreased by April to about 3. In May the comet had reached the limit of visibility with the naked eye, and nothing of the tail could be seen by this time.

David H. Levy succeeded in gaining the last visual sighting on 23 February 1988, almost two years after perihelion. The comet then had a magnitude of 17. Photographically, however, Comet Halley continued to be followed. In February and March 1993, a sudden magnitude outburst was observed, with the magnitude rising from 25 to 19. This event took place at a distance of 14.3 astronomical units.

In 2003, Halley's Comet had a magnitude of just 28. Three 8.2-metre telescopes, working together, were required to obtain this result.

Background and public reaction

As in 1910, the return of Halley's Comet was preceded by enormous interest in the media. Magazines put the comet on their covers and more than 100 books were published in various languages. Telescope businesses had a record turnover, with sales tripling. There were Halley medallions, Halley mugs, Halley stickers and much more. The hype before the appearance of the comet was considerable.

Comet Halley's return came at a time of enthusiasm for space. Investigation of the planets with space probes and the flights of the Space Shuttle generated a great deal of interest. With the approach of the comet, the excitement rose even higher. The European Space Agency launched a space probe called Giotto, which was to fly past the comet as closely as possible. In the event, the probe passed the comet's nucleus on 14 March at a distance of just 600 kilometres, and sent back the first direct images of a cometary nucleus. Halley's Comet proved to be a potato-shaped lump, about 10 kilometres across. The jets from the surface were clearly visible.

Japan sent the probes Sakigake and Susei. They passed the comet at significantly greater distances and measured the solar wind and interplanetary magnetic field. The Soviet Union had sent the probes Vega 1 and Vega 2. They flew past the comet on 6 and 9 March at distances of 8000 and 9000 kilometres, respectively. Comet Halley was observed and imaged in various spectral regions, and in addition there were magnetometers and dust-particle-impact detectors on board.

A lot happened on the ground as well. More than 1000 professional and 1200 amateur astronomers from 54 countries were part of the International Halley Watch. They exchanged observations and co-ordi-

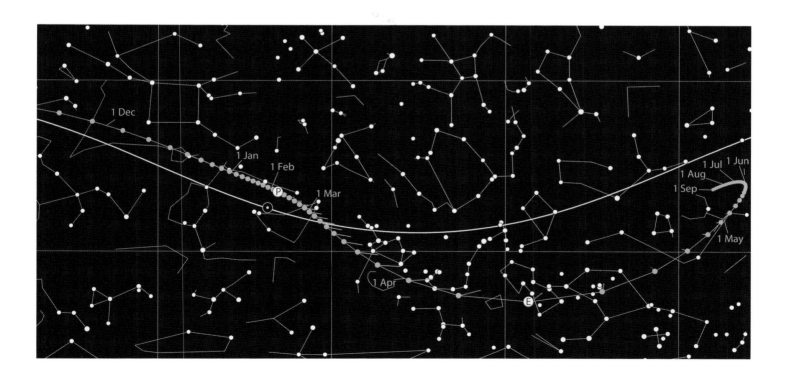

nated observing projects. In 1986, Comet Halley became one of the most observed comets in history.

For most of the general public, however, it was a different story. Admittedly there were, in the USA for example, enormous 'Comet Parties', such as the gathering in Jones Beach, New York, which drew as many as 40 000 visitors, who wanted to 'welcome' the comet. But most people were disappointed by the low brightness and the diffi-cult observing conditions. Many amateur astronomers deliberately cheated the latter by flying to the southern hemisphere, where they obtained many splendid photographs. For German-speaking ama-teur astronomers this was the start of astro-tourism in Namibia, sin-ce when every year astrophotographers, including those without any cometary interest, observe under the dark skies of south-west Africa.

The first direct images of a comet's nucleus caused a sensation in March 1986 with the fly-by of the European Giotto probe. Halley's Comet was found to be an irregular, potato-shaped lump 15 km × 8 km × 8 km. Only portions of the surface were active at the time of the fly-by (*bottom left*). Modern image-processing shows the overall view with the innermost tail cocoon (*below*), while the contour images (*bottom right*) reveal the outline.

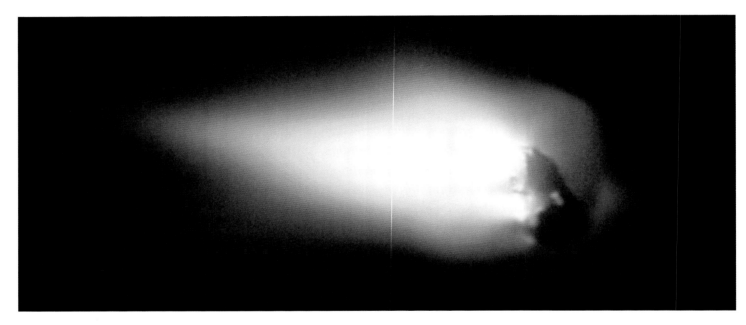

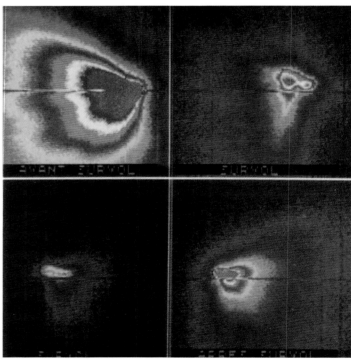

On 14 January 1986, the comet passed the bright galaxy Centaurus A (NGC 5128). *Bernd Koch*

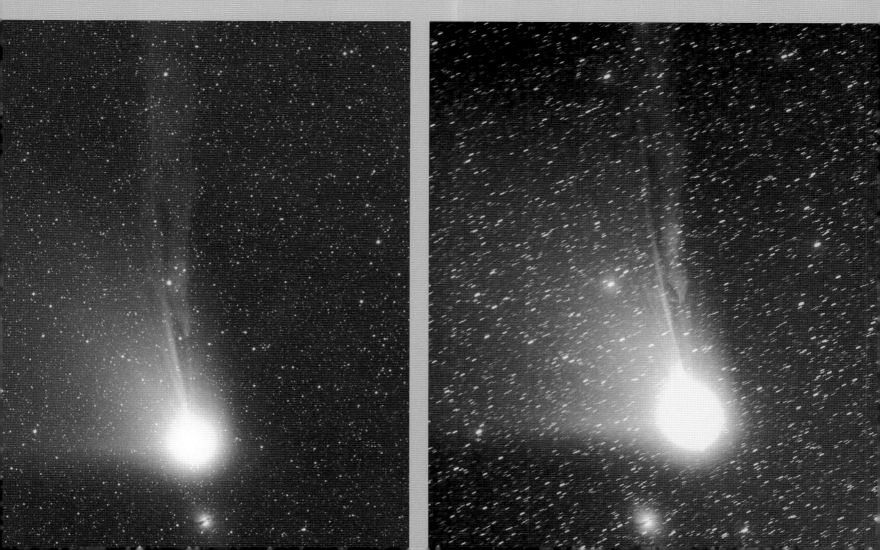

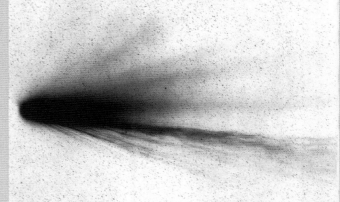

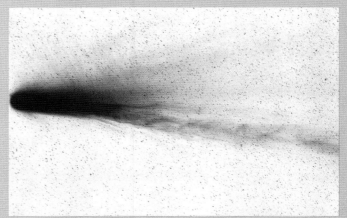

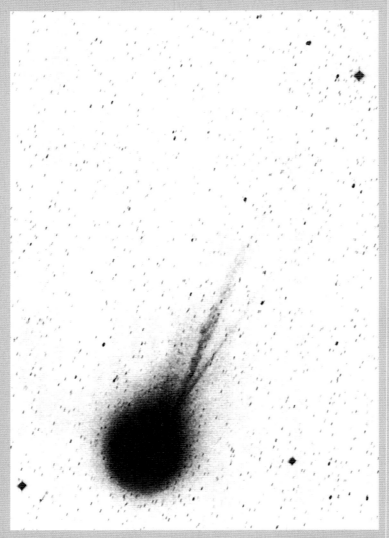

▲ The comet's gas tail on 9 December 1985, photographed with the European Southern Observatory's 1-m Schmidt telescope.

▶ The dynamic development of the tail, typical of Comet Halley, on 8, 9 and 10 March 1986. This sequence was also obtained at the ESO.

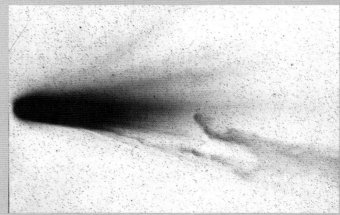

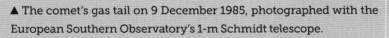

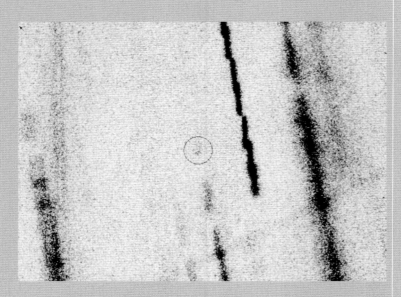

◀ Comet Halley on 10 January 1994, eight years after its major apparition. At a distance of 18.8 AU it is seen as no more than a tiny speck at the limits of detectability, between the star trails. CCD-image from the 3.6-m NTT (New Technology Telescope) at the European Southern Observatory.

Photographs through red, green and blue filters which show detail in the tail of the comet. The dust structures have a reddish tint, whereas the gaseous portion are blue.

☄ Comet Shoemaker-Levy 9 1994

This comet is an exotic among those in this book, because few people saw it themselves. Yet its effect on mankind and our understanding of motion in the Solar System has been enormous. For the first time the way in which two celestial bodies collided was actually observed. The comet, which had been broken into individual fragments by the tidal forces of Jupiter, hit the giant planet like a volley of shot. On Earth a wide section of the public watched, entranced, at what was then a unique spectacle.

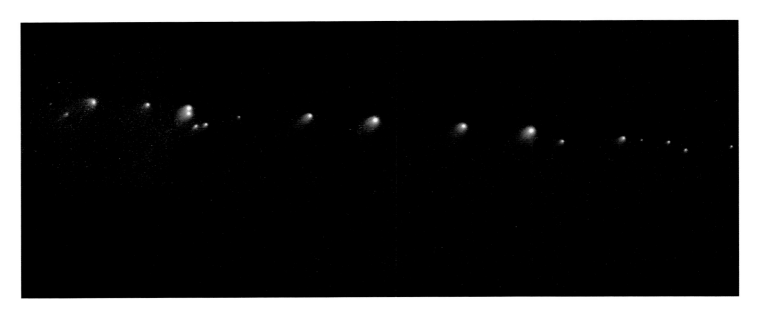

By any standards, Comet Shoemaker-Levy 9 was an extraordinary comet. Captured by Jupiter's gravity, which fractured it, a whole chain of comets orbited the largest planet in the Solar System.

Data	
Number:	27
Designation:	D/1993 F2 (Shoemaker-Levy)
Discovery date:	24 Mar 1993
Discoverer:	Carolyne and Eugene Shoemaker, David Levy
Perihelion date:	-
Perihelion distance:	5.3805 AU – 5.3794 AU
Closest Earth approach:	-
Minimum Earth distance:	-
Maximum magnitude:	12
Maximum tail length:	1'
Longitude of perihelion:	354.9° – 355.0°
Longitude of ascending node:	220.5° – 220.9°
Orbital inclination:	6.0° – 5.7°
Eccentricity:	0.9986

Orbit and visibility

D/1993 F3, generally known as Shoemaker-Levy 9 or, abbreviated, as SL9, was discovered as a satellite of Jupiter. It was the first discovery of a comet in orbit around a planet. By calculating the orbit backwards, it is now accepted that the comet had been orbiting Jupiter for 20 to 30 years. The gravitational force of the largest planet in the Solar System had diverted it from its original orbit and caused it to be 'captured' by Jupiter. The perihelion of the original orbit must have been in the asteroid belt between Mars and Jupiter, and aphelion somewhere near Jupiter's orbit. At one of the aphelion passages, Jupiter's gravity must have bound the comet to the giant planet. On 8 July 1992, a very close pass of the comet to Jupiter took place, when Shoemaker-Levy 9 grazed the planet, just 40 000 kilometres above the cloud deck. As it did so, the comet came under the influence of the strong tidal forces and was ripped into multiple fragments.

The resulting orbit made a collision with Jupiter unavoidable. Shoemaker-Levy 9 now reached a distance of no more than 50 million kilometres from the planet. The new orbital period amounted to two years, and led to a direct collision with the giant planet on its next pass of Jupiter. Between 16 and 22 July 1994, the fragments of the comet crashed into Jupiter at a velocity of 216 000 kilometres per hour.

Discovery and observations

D/1993 F2 was found by the most successful team of comet discoverers of the twentieth century. The Canadian author and science journalist David H. Levy had already made his name with numerous comet discoveries in the 1980s. The pair of scientists, Carolyn and Eugene

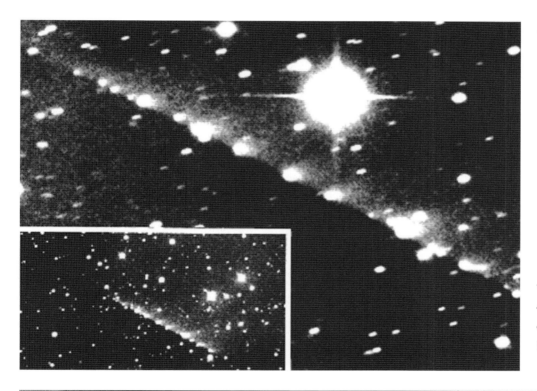

◄ After its discovery the 'string of pearls' formed by the fragments of the comet continued to grow in length.

▼ The comet as a fine line (just past the 4 o'clock position) at the beginning of July 1994, a few weeks before its impact with the giant planet.

Because the impacts took place on the far side of Jupiter, they were not visible directly from Earth. Only the Galileo Jupiter probe had a direct view of the events.

Shoemaker, had been engaged for more than ten years in the search for minor bodies in the Solar System.

The Shoemakers and Levy had planned to search for near-Earth small bodies on photographic plates that had been taken with the 460-mm (18-inch) Schmidt telescope at Mount Palomar. During a routine examination of the plates on 24 March 1993, Carolyn Shoemaker saw an object that looked 'like a squashed comet'. It was only 4° away from Jupiter and measured 50" × 10".

It was, by then, the eleventh comet discovery by the successful trio. The comet was subsequently found on other images back to 15 March 1993. It shape was unusual. The object looked like a string of pearls made from nebulous spots. The overall magnitude was only 14.

Better images just a few days later showed the fragmented comet as a chain of more than ten pieces. Later, as many as 21 comet fragments were identified and given the designations A to W (I and O were omitted). Over time, the portions of the comet drifted away from one another. In January 1994, the extent of the cometary chain was just 3', in June it had reached 8'. Many observers succeeded in visual sightings. The comet was even seen with a telescope aperture of just 150 mm. The magnitude amounted to 12 at the most.

Impact with Jupiter between 16 and 22 July 1994 took place on the farther side of the planet. The first impact was expected at 20:13 UT on 16 July. A gigantic fireball, 3000 kilometres in height, and lasting about half a minute was recorded by the Galileo space probe, which was in a 'favourable' position behind the planet. Further impacts occurred in the hours and days that followed, which were thus distributed over several of Jupiter's approximately 10-hour rotations. The impact sites at high southern latitudes overlapped to a certain extent.

Large, dark spots appeared in Jupiter's atmosphere at the impact sites, which reached sizes of up to 12 000 kilometres across. Sulphur compounds were the principal species detected, as well as iron, magnesium and silicon. In contrast, hardly any of the water expected from the comet was detected. The comet also triggered a temporary increase in polar auroral activity on Jupiter. The impact scars could even be seen clearly with small amateur telescopes – admittedly the latter were unable to see the comet itself, but its consequences could be experienced and were most impressive. After several weeks, the traces of the impacts on Jupiter's clouds faded and eventually completely disappeared. The comet did not leave any lasting changes in the circulation system on Jupiter.

Background and public reaction

Shoemaker-Levy 9 was a large comet. The original body had a diameter of about 5 kilometres. After the nucleus fractured, the overall dimension even increased. The more than 20 fragments with sizes between 200 metres and 2 kilometres stretched, before impact, over a length of 5 million kilometres.

It was already clear in the summer of 1993 that collision with Jupiter would occur a year later. As such, both science and the media had enough time to prepare for this great event. While astronomers got everything ready that the arsenal of space astronomy had to offer, it was, above all, the early Internet that contributed to the hype about the impacts. The Galileo Jupiter probe offered the possibility of directly observing the series of impacts. In addition, the new Hubble Space Telescope – whose capability had been restored to its initially planned state by a service mission in 1993 – followed the event. The results were made available on the Internet in almost real time – the comet impacts were one of the first Internet storms.

The first directly observed impact of one celestial body on another – with 600 times the strength of the world's total nuclear arsenal – did, however, renew anxieties about impacts on Earth. In subsequent years, numerous publications made use of this fear, and capitalized on the forthcoming end of the millennium and the resulting wave of mysticism.

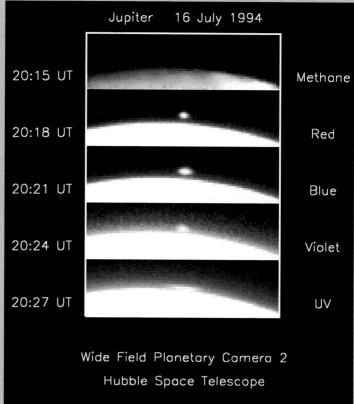

Jupiter   16 July 1994

20:15 UT	Methane
20:18 UT	Red
20:21 UT	Blue
20:24 UT	Violet
20:27 UT	UV

Wide Field Planetary Camera 2
Hubble Space Telescope

▲ The Hubble Space Telescope was able to obtain images of the impacts at various wavelengths.

▲ After they had rotated onto the facing side of Jupiter, the impact sites showed as striking dark spots.

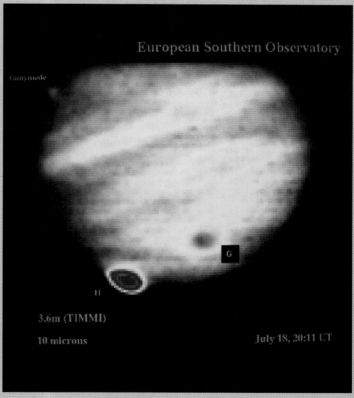

European Southern Observatory

Ganymede

G

H

3.6m (TIMMI)

10 microns   July 18, 20:11 UT

▲ The impacts were seen more clearly in the far infrared than in the visual spectral region. This image was obtained by ESO's 3.6-m telescope at La Silla in Chile.

☄ Comet Hyakutake 1996

As a phenomenon in the second half of the twentieth century it was unsurpassed. Hyakutake took everyone who was lucky enough to see it during the days when it was closest to Earth in March 1996 by surprise. The bluish comet gleaming directly overhead, and with a gigantic tail that spanned the sky above an icy winter landscape, remains an unforgettable experience. The almost ominous impression resembled the broadsheets of former centuries: so mighty may the phenomenon of a comet appear.

A comet par excellence. A detailed image of Comet Hyakutake. *Gerald Rhemann*

Data

Number:	28
Designation:	**C/1996 B2 Hyakutake**
Discovery date:	**30 Jan 1996**
Discoverer:	**Yuji Hyakutake**
Perihelion date:	**1 May 1996**
Perihelion distance:	**0.23 AU**
Closest Earth approach:	**25 Mar 1996**
Minimum Earth distance:	**0.10 AU**
Maximum magnitude:	**−0.8**
Maximum tail length:	**100°**
Longitude of perihelion:	**130.2°**
Longitude of ascending node:	**188.1°**
Orbital inclination:	**124.9°**
Eccentricity:	**0.9997730**

Orbit and visibility

Hyakutake's appearance in 1996 was not its first visit to the Sun. Before this apparition, the comet had an orbital period of 15 000 years with aphelion at a distance of 1300 astronomical units. Altered by planetary perturbations during its perihelion passage in 1996, the orbit was elongated with aphelion now at 3500 astronomical units from the Sun. Comet Hyakutake will visit us again in 72 000 years.

Hyakutake's visibility was favoured by its relatively small perihelion distance of about one quarter of the Earth's distance from the Sun, as well as by its small separation from the Earth of a few million kilometres. Because its closest approach to Earth occurred significantly before perihelion, the comet's path took it far from the Sun into the dark night sky.

At its discovery C/1996 B2 was in the constellation of Libra, in which it remained until 20 March 1996. From the third week of March, it began to change its position towards the north. The comet crossed the constellations of Boötes and Draco and finally arrived near the North Celestial Pole in the constellation of Ursa Minor.

Two days after its closest approach to Earth, on 27 March the comet passed just 4° from the Pole Star. As such, for observers at northern locations, it was high in the night sky at the period of greatest brightness. Nei-

ther moonlight nor twilight interfered, so that Hyakutake could be observed throughout the night.

In the last days of March, C/1996 B2 travelled south along the boundary between Cassiopeia and Camelopardalis. At the beginning of April it passed into Perseus, and at the end of the month it encountered the Pleiades and Venus in Taurus in the evening sky. The following passages past Mercury and the crescent Moon were better seen from southern locations.

After perihelion on 1 May, in which the comet was very close to the Sun and could not be seen, it plunged into a southern declination in the constellation of Cetus after 9 May. Moving away from both the Sun and the Earth, during the next few weeks the comet passed through the constellation of Eridanus and even deeper into the southern sky. The last observations were reported when it was in the constellation of Triangulum Australe near the southern Milky Way.

Discovery and observations

On the morning of 30 January 1996, the Japanese amateur astronomer Yuji Hyakutake scanned the sky above his southern Japan home with giant 25×150 Fujinon binoculars. He scrutinized a point in the constellation of Libra, where, a month before, he had discovered his first comet, C/1995 Y1. Only 3° distant from the position of his first triumph, Hyakutake once again found an unknown nebulous patch.

At discovery, Hyakutake's second comet was magnitude 11 and 2.5' across. Later, the comet was found on a photograph taken before its discovery on 1 January by the Japanese astronomer Kesao Takamizawa, with a magnitude of 13, and 1' in extent. C/1996 B2 became continuously brighter and larger. At the beginning of February its magnitude had already climbed to 9, and it was 7' in size. Shortly afterwards, the comet had already progressed to being visible in bino-

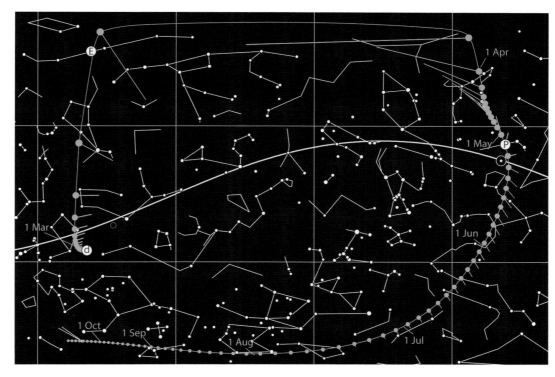

culars, and by the end of the month, at a magnitude of 6, it could be seen with the naked eye under dark skies. Through a telescope, a tail, 1° in length could already be seen.

In the middle of March, after the Moon had left the sky, the most spectacular phase of visibility began. On 19 March, C/1996 B2, already at magnitude 2.5 was visible with a significant tail. In the following days, the comet became recognizably brighter by the day, and moved ever faster. The 0 magnitude mark was passed on 24 March, and at its closest approach to Earth on 25 March, the maximum value of −0.8 was reached.

The comet presented a magnificent sight at this stage. The coma, about 1.5° across, shone with a bluish-green light, and from it sprang a bluish gas tail, which was continually getting longer. On 23 March it measured 45° long, on 25 March it was seen to be between 50° and 80°, and on 26 March it reached a length of 90°. For two or three nights there was the overwhelming sight of the comet apparently stretching right across the sky.

At closest approach to Earth, the comet raced across the sky with an apparent velocity of 1° per hour and changed its position significantly during the course of a single night. Spiral-shaped jets near the nucleus were visible, even through binoculars. Spectacular structures accompanied a tail disconnection event on 24 March. The separated portion of the tail was clearly visible with the naked eye as knots in the tail. Similar events had already occurred on 28 February and 10 March.

On 27 March the magnitude of the comet was about 0.5, and the tail was seen to a length of between 45° and 70°. On 30 March this value had dropped to 30°, and the magnitude was down to 1.5. Until the middle of April the magnitude sank even lower to below 2.5, but however rose again after perihelion. At this period, the tail appeared to most observers at a length of between 10° and 15°. In April, the previously observed gas tail was joined by a faint dust tail, about 5° long.

On 15 April there was a striking magnitude outburst. At the end of April, C/1996 B2 again reached a magnitude of 2. The comet could be followed until the 28 April, when it was 12° from the Sun – this ended its visibility for European observers. Passage past the Sun could be followed over the Internet, thanks to the SOHO solar observational satellite that had recently been launched into space. This showed the tail to be triple, and the magnitude of the comet had again reached 0.

After perihelion, the comet was seen from southern locations until 9 May, at magnitude 3. In comparison with the head, the tail was now significantly brighter than before perihelion. In the middle of May it still exhibited a length of 4°.

Until the end of June 1996, Hyakutake's Great Comet could be followed by the naked eye. On 1 July its magnitude still amounted to 7, but by 1 September it had dropped to 11. The last photographic image in October 1996 showed it at magnitude 17.

Background and public reaction

Comet Hyakutake's apparition was marked by its sudden appearance and the short period of its high point of about a week at the time of closest approach to Earth. The small distance of only 40 times further than Moon and the path of its orbit away from the twilight zone led to the comet's great brilliance. Bad weather in Central Europe largely prevented any widespread effect on the general population. Another contribution was the low surface brightness of the tail, which only displayed its enormous length under dark skies, but which was hardly visible in light-polluted areas.

C/1996 B2 was, with the diameter of its nucleus less than 3 kilometres, rather a small comet. Its activity was above average, as the jets and tail disconnection events that were recorded by the Hubble Space Telescope, showed. Comet Hyakutake emitted up to 10 tonnes of material per second, with a velocity of up to 500 metres per second. The blue light was primarily caused by C2-emission and ionized carbon monoxide. In addition, the comet contained large fractions of methane and ethane – possible evidence of the chemical composition of the early stages of the Solar System.

One novelty was the X-ray emission discovered with the ROSAT satellite, and which probably arose from the interaction of the comet with the solar wind. The tail was 80 million kilometres long in April. The nucleus rotated on its axis with a period of 6 hours 13 minutes.

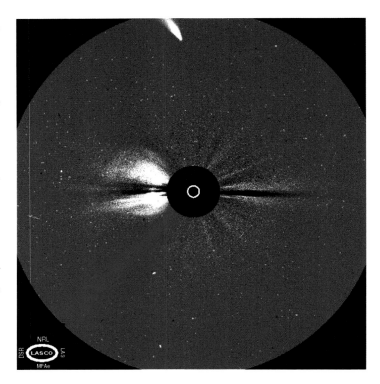

The perihelion passage of the comet at the beginning of May 1996 could be followed on images from the SOHO solar probe.

The dynamical changes in the nucleus were one of the outstanding properties of Comet Hyakutake. *Tony und Daphne Hallas*

The straight tail, stretching across half the sky was particularly impressive from dark locations. *Stefan Binnewies, Peter Riepe (top), Uwe Wohlrab (below left), Gerald Rhemann (below right)*

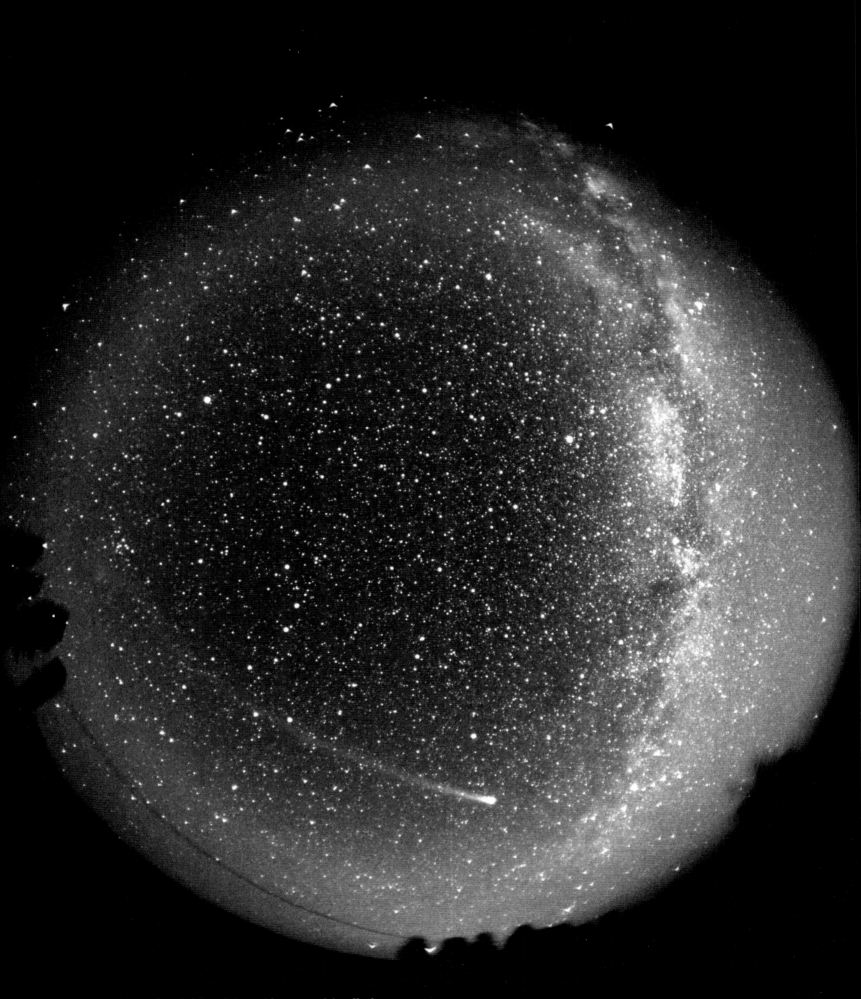

The enormous length of the tail is evident on this all-sky
photograph on 28 March 1996. *Michael Kobusch*

Tail disconnection events were a characteristic feature of Comet Hyakutake, particularly the event on 24–25 March (colour images). *Bernd Liebscher (above), Jerry Lodriguss (below), Nornbert Mrozek (top left), Gerald Rhemann (centre left and bottom left)*

These colour images show the gas and dust portions of the tail at the beginning (*above*) and the end (*below*) of April 1996. *Gerald Rhemann*

Comet Hale-Bopp 1997

Never before have so many people followed a comet for such a long time. Hale-Bopp was visible in the sky to the naked eye for 17 months. At the time of its greatest brightness for several weeks it was visible the whole night from Central Europe and, with its bright gas and dust tails, it offered a perfect example of a spectacular comet. Comet Hale-Bopp was also a record in other respects. It possessed a nucleus that was five times the size of Comet Halley's, and it produced a large quantity of dust and gas, hardly equalled by any other comet before it. Eleven years after Comet Halley, Comet Hale-Bopp fulfilled what the former comet had merely promised.

Comet Hale-Bopp was impressive because of its bright dust tail and the structures within the gas tail, as seen here on 2 April 1997.

Data

Number:	29
Designation:	C/1995 O1 Hale-Bopp
Discovery date:	23 Jul 1995
Discoverers:	Alan Hale, Thomas Bopp
Perihelion date:	1 Apr 1997
Perihelion distance:	0.914 AU
Closest Earth approach:	22 Mar 1997
Minimum Earth distance:	1.31 AU
Maximum magnitude:	−0.7
Maximum tail length:	25°
Longitude of perihelion:	130.6°
Longitude of ascending node:	282.5°
Orbital inclination:	89.4°
Eccentricity:	0.9951279

Orbit and visibility

Comet Hale-Bopp did not make its first visit to the Sun's neighbourhood in 1997. Its orbit, almost at right-angles to the Earth's orbit had brought it into the inner Solar System 4265 years previously. Because of a near approach to Jupiter on 5 April 1996 the orbit was greatly altered, and the next return is expected after 2404 years, in the year 4401. At the same time, the orbit shrank, with aphelion distance being reduced from 525 to 368 astronomical units. Comet Hale-Bopp has undergone repeated perturbations by the planets in the past.

C/1995 O1 was, at the time of its discovery, in the constellation of Sagittarius near the Milky Way. Because it was discovered at a very great distance from the Sun, it hardly moved in the first months, and only the motion of the Earth around the Sun was mirrored in the comet's orbital

loop in 1995. In November 1995 the first of numerous periods of visibility of the comet came to an end, as it faded into the evening twilight.

At the beginning of January 1996, as seen from Earth, the Sun passed just 2° away from Comet Hale-Bopp. In the following summer, the comet was again to be seen in the constellation of Sagittarius. In November 1996 the second period of visibility came to an end.

On 3 January 1997, there was the second solar conjunction, this time at a distance giving a separation of 27°, because the comet was north of the Sun in the constellation of Aquila. Now the best visibility of C/1995 O1 followed. Initially in the morning sky in the constellation of Sagitta, the comet then passed during March and April 1997 along the autumn Milky Way from Sagitta into Andromeda. In doing so, it moved to ever greater northern declinations, and for locations with latitudes of 45° north or greater it became circumpolar, and thus visible throughout the night. So closest approach to Earth on 22 March, as well as perihelion on 1 April, occurred under extremely favourable conditions for Central Europe. The Moon interfered with observations only at the end of March and the end of April.

At the end of March, Comet Hale-Bopp passed the Andromeda Galaxy, M31, at a distance of just 5° – a much-photographed event. In April the comet again moved south and crossed the constellations of Perseus, Taurus and Orion to its third conjunction with the Sun. In the middle of May it was last seen from Europe.

In June, Comet Hale-Bopp moved to southern declinations, so after solar conjunction the comet could be seen only in the southern sky. Located in the constellation of Puppis from August to October 1997, it had one last period of unfavourable morning visibility for observers in the northern hemisphere, whereas farther south, the comet could be easily observed. At the end of the year it had reached the constellation of Dorado, so that the view of the comet as it retreated from the Sun and the Earth, in the years following, was reserved for observers deep in the southern hemisphere.

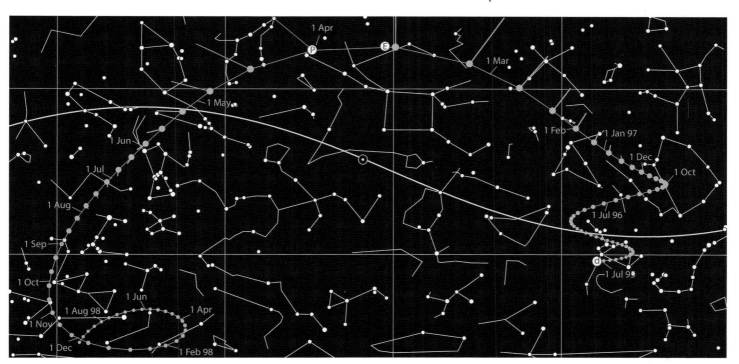

Discovery and observations

Comet Hale-Bopp is one of the most wonderful examples of how amateur astronomers may, by accident, become discoverers. In this case it was two American observers who found the comet independently of each other. On the night of the 23 to 24 July 1995, both, by chance, had their telescopes pointed at M70, one of the rather unimpressive globular clusters in the constellation of Sagittarius.

Alan Hale had already been considerably involved in the search for comets. He had monitored the sky for comets for 400 hours, but had given up searching, and had instead concentrated on following comets that had already been discovered. As such, he had already observed over 200 comets. On 23 July another was to be followed. Hale had got his 16-inch Dobsonian ready. But his target object was, however, too low on the horizon, so Hale intended to pass the time by looking at M70.

On the same night, Thomas Bopp also pointed his telescope at M70. It was not even his own telescope – Bopp was observing clusters and galaxies with a group of friends using a 17.5-inch Dobsonian. Bopp had not seen a comet before – until that night.

Both amateurs noticed a nebulous patch, at magnitude 10.5, and 2' across, not far from M70, that was not on their star charts. Over the course of the night, Hale detected a slight change in position, and was the first to report the comet. He thus became the first in the list of names of the discoverers.

The comet could already have been discovered significantly earlier. An image by Robert McNaught from Siding Spring Observatory showed it an unbelievable four years earlier on 27 April 1993, at a brightness of 18 and 20" in extent.

By November 1995, the brightness had risen slightly to magnitude 10, and a tail, 10' long, had formed. Because of the comet's enormous distance, still far beyond Jupiter's orbit, it was obvious that either a very large or very active comet must be involved. Five smaller outbursts of brightness in 1995 confirmed this.

In January 1996 after its reappearance from the glare of the Sun, C/1995 O1 was observed at a magnitude of 9. This value increased throughout the course of 1996, to magnitude 8 in April and 6.5 in May. In spring 1996, Comet Hale-Bopp could be seen even with binoculars, and by May sightings with the naked eye were reported.

On 7 June, Comet 22/P Kopff passed the field of C/1995 O1, moving within 3° of the latter. The periodic comet was then in the background, and appeared fainter than Comet Hale-Bopp, which was moving towards the Sun. Both could be seen in binoculars at the same time.

From June to September 1996, the brightness of Comet Hale-Bopp stagnated. If its approach is taken into account, there was actually a decline of 0.3 magnitudes. Evidently the comet went through a faint phase, and for some time it was questionable whether the great expectations for the period around perihelion would be fulfilled. Comet Hale-Bopp finished the year with a magnitude of 4 at the beginning of December.

The first observations after the second solar conjunction banished any doubt. On 20 January 1997, Comet Hale-Bopp showed a magnitude of 2.5, and a tail, 2° long was visible. By the beginning of February, the magnitude had risen to 2. By the end of February it had already reached 1, and a magnitude of 0 was reported at the beginning of March. The length of the tail rose from 2° to 10°. During the total solar eclipse of 9 March 1997, the comet was already as bright as the brightest stars and was visible with the naked eye whilst the Sun was eclipsed.

From the end of March to the beginning of April 1997, Comet Hale-Bopp was a spectacular sight. Its magnitude reached –0.7 during this period. With well-separated gas and dust tails it was in the sky the whole night long, although it was best observed in the evening and in the morning. The dust tail, with its high surface brightness, was clearly seen from towns. Its length reached 20°. The significantly fainter gas tail, however, required dark skies. Visually, it appeared whitish, and the striking blue colour seen on photographs was hardly detectable.

At this period the sight of the comet's head was also spectacular. A coma, 15' in diameter surrounded the bright nucleus. Telescopically, with medium to high magnifications, numerous bright semicircular shells could be seen around the latter. These are known as envelopes and are formed from material ejected from rotating jets. These jets eject material only when they lie on the side of the cometary nucleus that is turned towards the Sun. The fountains then produce a ring at each rotation, so the structure of the envelopes changes from one evening to the next.

During the weeks of best visibility, striae and synchrones developed in the dust tail as a result of the activity taking place at the nucleus, and which resembled similar phenomena seen in Comet West. In the meantime, the gas tail became wider and fainter. At the end of April it still had a length of 10°.

From April into May, the magnitude of the comet faded, reaching just 2 at the end of May 1997. The length of the tail shortened to just 1° in June. After solar passage, Comet Hale-Bopp was seen again at a magnitude of 3. The length of the tail was 0.5°. By the beginning of December, the magnitude of the comet fell to 7. The last observations without optical aid are recorded from 9 December 1997.

In subsequent years, the comet could be followed as its magnitude declined. In February 1998, its magnitude was still around 8 and, until the end of the year, remained within the reach of binoculars. In February 2003 it was still magnitude 16. Until about 2020 with current professional astronomical methods, it should still remain possible to follow Comet Hale-Bopp on its way into the outer Solar System.

Background and public reaction

Even the circumstances suggested that Comet Hale-Bopp would be an unusual comet. The visual discovery by Hale and Bopp came at a time when the comet was still at an enormous distance of 7.2 astronomical units. That corresponds to a position between the orbits of Jupiter and Saturn. This is the greatest distance at which a comet had ever been discovered visually. McNaught's pre-discovery image would have stretched the distance to a tremendous value of 13 astronomical units, had it only been recognized at the appropriate time.

With discovery two years before the expected spectacle, both science and the media had enough warning to get ready for the comet. The comet's faint phase in 1996, was largely ignored by the media, despite many prophesies of doom (dubbed 'Hale-Flop'), and the real hype about the comet began in March.

A major role in encouraging awareness among the general public was the fact that the comet was conspicuous in the night sky for four months between February and May – much longer than most comets, which are clearly visible for just a few days or weeks. In addition, the comet's favourable position for northern, heavily populated countries on Earth was decisive – never before had so many people been able to see a comet. With a total of 569 days of visibility, Comet Hale-Bopp had the longest period of visibility with the naked eye of any comet in history.

Even at the time of its discovery, the enormous size of the comet was clear. At a distance when the outgassing of many comets cannot even be detected, the coma was already three million kilometres in diameter. The diameter of Comet Hale-Bopp's nucleus was determined by the Hubble Space Telescope to be 40 kilometres in diameter – nearly five times the size of Comet Halley. The nucleus rotated with a period of 11 hours, 20 minutes.

At the peak of its activity, C/1995 O1 lost 400 tonnes of dust and 300 tonnes of water every second – 100 times the amounts from Comet Halley. Nevertheless, Comet Hale-Bopp lost only about 0.1 per cent of its mass in its encounter with the Sun.

Comet Hale-Bopp was a gas-rich comet. It emitted about two to five times as much in the form of gas as in dust. Carbon monoxide and water vapour were the primary gases detected. Between the gas and dust tails a third tail of neutral sodium, 28 million kilometres long, was also observed. Comet Hale-Bopp also revealed numerous chemical compounds, such as sulphur dioxide, previously unknown in comets.

The long preparation time also paid off when it came to scientific research. Apart from the Hubble Space Telescope and numerous professional observatories, the astronauts on board the Space Shuttle Discovery were able to observe the comet in the middle of August.

An American sect contributed a tragic and rather macabre chapter to the history of the Great Comet. Marshall Applewhite and Bonnie Nettles founded a UFO religion at the end of the 1970s that was moulded on a crude, esoteric conception of the world, and a life of asceticism, compulsion and control of the members, who lived commu-

nally. When the American amateur astronomer, Chuck Shramek, took a digital image of Comet Hale-Bopp in November 1966, he asserted that he had photographed a 'Saturn-like object' in the tail, that was not contained in his star-chart software. It was rapidly established that Shramek had incorrectly configured his star-charting program, and that the Saturn-like excrescence could be explained as perfectly ordinary background stars, affected by diffraction in the optics used. Nevertheless Shramek spread his alleged finding countrywide through a popular radio show, so that it was readily taken up by other conspiracy theorists. Finally, the Internet was full of claims that a UFO was heading for Earth in the tail of the comet, but that this was knowingly concealed by NASA and scientists.

For the members of the Heaven's Gate sect this was the long-awaited signal to 'leave' the Earth. In March, the sect's founder, Applewhite, announced a group suicide on the sect's Internet site. On 26 March, 39 corpses, including that of the founder, were found at the sect's property near San Diego, California.

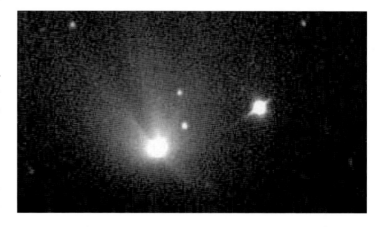

The photograph by Chuck Shramek (*top*) with the image of the allegedly 'Saturn-like' object (the star with the diffraction spikes on the right) alongside the comet, that unleashed the UFO hysteria surrounding Comet Hale-Bopp. It ended with the mass suicide by the Heaven's Gate sect.

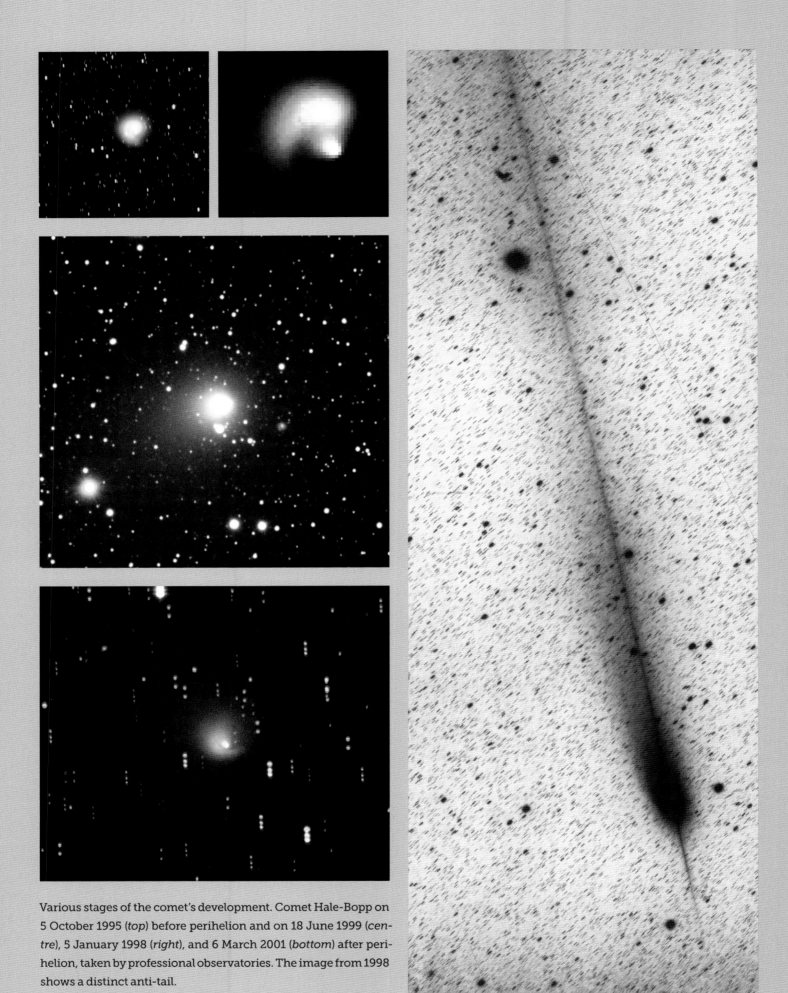

Various stages of the comet's development. Comet Hale-Bopp on 5 October 1995 (*top*) before perihelion and on 18 June 1999 (*centre*), 5 January 1998 (*right*), and 6 March 2001 (*bottom*) after perihelion, taken by professional observatories. The image from 1998 shows a distinct anti-tail.

Comet Hale-Bopp in April 1996, photographed from Tau-
tenberg Observatory. The tail structures were enhanced by
unsharp masking.

At the time of greatest brightness, the dust tail could be clearly seen in twilight or even over cities. *Bernd Liebscher (above), Christoph Ries (below)*

▲ At dark locations, the gas and dust tails could be clearly distinguished. *Pekka Parviainen*

▼ On 6 April 1996 the comet passed the Double Cluster h and χ Persei, in the constellation of Perseus. *Uwe Wohlrab*

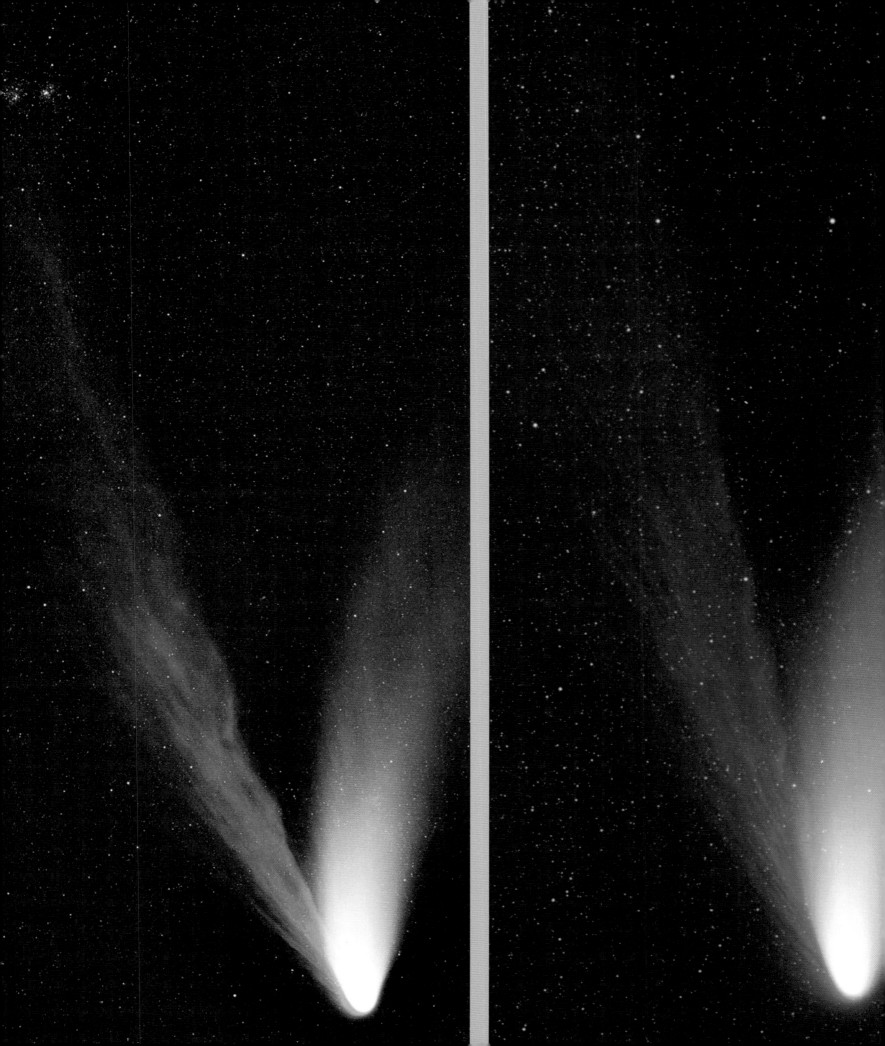

◀ There is an impressive difference in colour between the blue gas and white dust tails on 1 April 1997 *Gerald Rhemann (opposite left), Norbert Mrozek (opposite right)*

Comet Hale-Bopp over the northern horizon on 1 April 1997 (*above*) and 27 March 1997 (*below*). *Stefan Binnewies, Bernd Schröter (above), Gerald Rhemann (below)*

Long exposures of the tail structures on 7 March 1997 (*above*) and 8 April 1997 (*below*). *Philipp Keller*

Drawings of the tail structures on 1 April 1997 (*above*) and 8 April 1997 (*below*), using a 14-inch telescope. *Ronald Stoyan*

▲ The asymmetrical distribution of brightness is clearly shown in this photograph of 10 March 1997. The brightest region is marked by the shells, known as envelopes. *Bernd Koch*

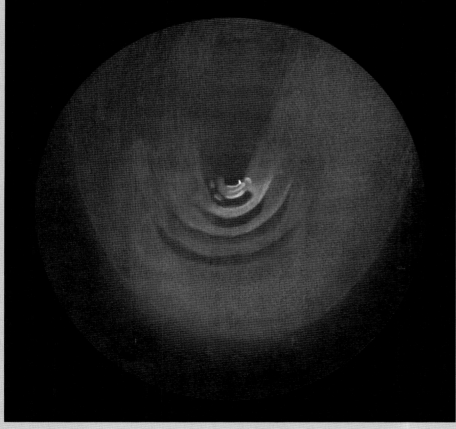

◄ Drawing of the nuclear region on 15 April 1997 through a 14-inch telescope. The semicircular shells or envelopes become brighter towards the almost star-like false nucleus. *Ronald Stoyan*

☄ Comet McNaught 2007

The last great comet to date appeared on the celestial stage in 2006, but only became a major spectacle in January 2007. Even though it was visible for just a few days in daylight, it is among the brightest comets ever observed. The impressive tail, the rays of which resembled those of the comets of 1744 or 1976, first displayed its full splendour after perihelion and only for the inhabitants of the southern hemisphere. For Europeans there was only a pale echo of that splendour in the rather ghostly glow of the tips of the tail above the winter horizon.

Comet McNaught setting two days before perihelion on 10 January 2007 above the Alpine peaks. *Rudolf Dobesberger*

Data	
Number:	30
Designation:	C/2006 P1 McNaught
Discovery date:	7 Aug 2006
Discoverer:	Robert McNaught
Perihelion date:	12 Jan 2007
Perihelion distance:	0.171 AU
Closest Earth approach:	15 Jan 2007
Minimum Earth distance:	0.81 AU
Maximum magnitude:	−5.5
Maximum tail length:	35°
Longitude of perihelion:	156.0°
Longitude of ascending node:	267.4°
Orbital inclination:	77.8°
Eccentricity:	1.0000183

Orbit and visibility

C/2006 P1 was discovered on 2006 when it was opposite the Sun in the constellation of Ophiuchus, near the summer Milky Way. The first orbital calculations showed a perihelion distance comparable with that of the planet Mars, but further observations soon revealed that the comet would follow an orbit with perihelion at just 0.17 astronomical units. With this perihelion, significantly inside the orbit of Mercury, a very bright cometary apparition could be expected.

In September 2006, Comet McNaught moved to the borders of the constellation of Scorpius. At this southern location it was difficult for observers in the north to catch sight of the comet. In October and November the visibility conditions deteriorated for observers in the

southern hemisphere as well, because it was getting closer to the Sun. Finally, the first phase of visibility came to an end in the evening twilight in the middle of November 2006.

At the end of December 2006, the comet could again be seen, in the morning sky, a very short distance from the Sun. For European observers, however, unfavourable observing conditions near the horizon persisted, because of the comet's southern location. In the second week of January there was a second visibility window, low in the western evening sky. Despite the small distance from the Sun of less than 15° during the whole of this second phase of visibility, the comet was easier to see, because of its rapidly increasing magnitude.

After 10 January, the comet neared the Sun. After perihelion on 12 January, there was a solar conjunction on 14 January, at a distance of 5°. Forward scattering of sunlight by the particles in the tail increased the brightness of the comet considerably, so that on 14 January it reached its maximum magnitude of about −5.5. On 15 January, the comet was closest to the Earth. The distance, at about 0.8 of the distance from Earth to the Sun, was slightly closer than that of Comet Hale-Bopp ten years previously.

After 14 January, the comet moved into the southern sky, and in the third phase of visibility, the comet remained unobservable for European observers. After 15 January, inhabitants of the southern hemisphere could marvel at it in the south-western evening sky during the period when it was at its greatest brightness. The comet remained an evening object in the southern sky until July 2007, when it was last seen in the constellation of Musca.

The orbital period of C/2006 P1 is thought to be 92 000 years, with the comet being significantly accelerated by its passage past the Sun in 2007.

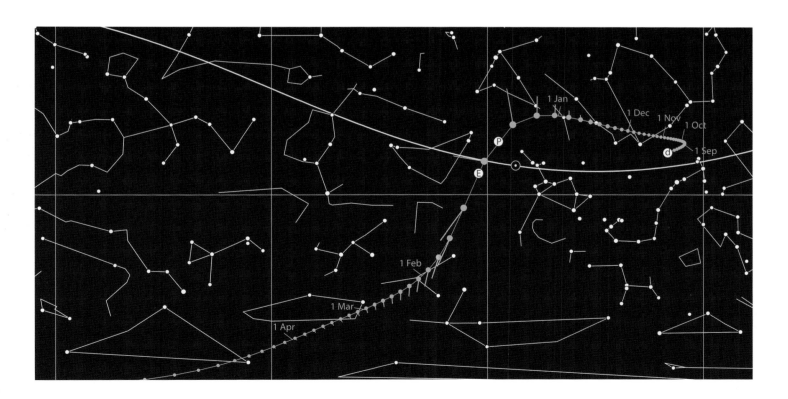

Discovery and observations

C/2006 P1 was the 35th comet discovery by the Scottish-Australian professional astronomer Robert McNaught and the 31st to carry his name. Like most of these comets, C2006 P1 was found photographically on images taken for the Siding Spring Survey, an automatic monitoring of the southern sky for near-Earth asteroids (NEAs). It was carried out at the observatory of the same name, about 400 kilometres northwest of Sydney.

At discovery the comet was the faint magnitude of 17.3, and hardly 20' across. Its distance from the Sun at discovery was more than 3.5 times the Earth's distance.

Initially there were fears that McNaught's promising discovery was too faint to develop into a truly great comet when near the Sun. Two months after discovery it was still only magnitude 12, and no more than 5' across. By the end of the first phase of visibility in November 2006, the brightness had only risen to about magnitude 9.

On its return to the sky before Christmas 2006, the magnitude had, however, already risen to 4, and it size had increased to 0.5° across. As such, the comet could now be seen, under good conditions, with binoculars. The magnitude now rose rapidly to about 1.5 on 5 January. The nucleus in particular became significantly brighter and a tail formed, about 1° long. Because of the difficult observing conditions with the comet being close to the horizon and also near the Sun it was a difficult object for naked-eye observation, and was only occasionally detected.

In the second week of January the rate of brightness increase itself rose. The comet now brightened at 1 magnitude per day. As such, assuming an appropriate view of the horizon, it became a significant object in the bright morning or evening twilight. Most of the observations from Europe were made during this period, with the comet often seen low over hills or towns.

On 13 and 14 January C/2006 P1 reached its peak magnitude at –5.5. It then was close to the Sun and for almost the whole world, was only visible at the same time as the latter. But for high northern latitudes there was just the possibility of seeing the comet for a short time in the very bright twilight. Numerous observers reported successful sightings of the comet in the daytime sky. Between 12 and 14 January it was easy to see with binoculars and could also be made out with the naked eye, if the Sun was hidden. Frequently it was taken for a condensation trail, because its short, comma-like shape greatly resembled one.

Up to then the comet had remained an object for specialists, who could find it near the Sun with optical aids, and could exercise the necessary precautions in doing so. The following impressive phase, which earned Comet McNaught its place among the great comets, was reserved primarily for observers in the Earth's southern hemisphere.

At a magnitude of about –3, the comet dived back into the southern sky – sufficiently to make it immediately visible to the naked eye. In doing so, it developed a spectacular tail that changed from evening to evening. On 17 January it was detected at a length of 3°, one day later it was already 15°, on 20 January, 20° and on 23 January, 35°. The wide, fan-like tail resembled the appearance of Comet

West, 31 years previously. The numerous tail rays (striae), produced the impression that there "were eight to ten comets simultaneously visible", as the American observer Stephen James O'Meara noted on 20 January from Hawaii.

Unique in the history of the appearance of the great comets as far as many observers were concerned was, above all, the 90° arc described by the tail. As such, the northernmost portions of the tail were even visible above the horizon from Europe, while the comet's head was invisible below it. The last time such a phenomenon had appeared in the European sky was with the Great Comet of 1744.

Nevertheless, the lavish splendour soon waned. At the end of January, the magnitude of the comet had already dropped to 2. At the beginning of March it was only 6. While the tail remained clearly visible at the same time, still with a length of 10°, the head of the comet could no longer be seen with the naked eye. In February, C/2006 P1 formed an anti-tail, clearly visible telescopically, that was 12' long, which could be attributed to the sight geometry and to the particles, released by the comet, that lay between it and the Sun. By the middle of 2007 the magnitude had finally declined to 10. The last observations were from the Australian Kambah Observatory on 11 July 2007.

Background and public reaction

The phenomenon of the Great Comet of 2007, and thus its effects, were confined to the general public only in the Earth's southern hemisphere. In Europe and the USA, the brightest phase occurred in the daytime sky or in deep twilight. It was, therefore, largely ignored by the media. Numerous observers travelled south after the comet as it was vanishing southwards following perihelion in the middle of January. This was how the impressive photographs were obtained by American and European photographers from locations such as Chile and Namibia.

Comet McNaught was the first great comet that was primarily followed by amateurs using CCD-cameras. This technology, restricted to professional astronomers at the time of the apparition of Comet Halley, had developed greatly and replaced the former chemically based photography that was still used for Comet Hale-Bopp in 1997. By exchanging digital images via the Internet, it was possible for members of the amateur comet scene to rapidly exchange information on the position and magnitude of the comet.

Extensive professional research was carried out by ground-based observatories. The solar observatory SOHO was able to record the passage of the comet close to the Sun – and the event could be followed live on the Internet. Rather unintentionally, the Ulysses probe was caught in the tail of the comet on 2 February. As a result, O_3^+ was detected in a comet's tail for the first time. Moreover, researchers were surprised at the strong influence of the comet on the solar wind, whose velocity was almost halved, although the nucleus of the comet was at a distance of 100 million kilometres. C/2006 P1 possessed a large fraction of dust. A plasma tail could not be observed but, as with Comet Hale-Bopp, tails of neutral iron and sodium, which appeared particularly strongly in the spectrum, were both detected. The tail reached its maximum extent on 19 January, with a length of 150 million kilometres, and was 65 million kilometres wide.

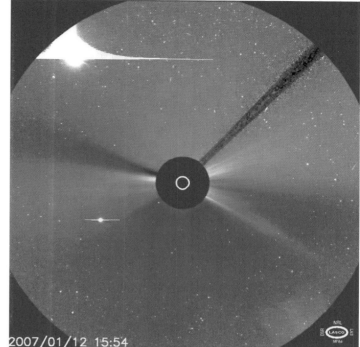

2007/01/12 15:54

► At perihelion on 12 January 2007 the comet entered the field of view of the SOHO solar probe.

▼ The comet in bright twilight on 10 January 2007 above Graz in Austria. *Burkhard Leitner*

Typical views of the comet in the middle of January at perihelion. Because of the close vicinity of the Sun, observations were possible only in twilight or in the daytime sky. *Unknown photographer*

▲ By 17 January 2007, the tips of the comet's tail stretched into the sky above Cerro Paranal in Chile. *E. Jehin and A. Correa*

▼ Until 22 January 1997 it was also seen in Europe, despite the Moon's glare. *Stefan Binnewies*

▲ On the evening of 22 January 2007 the comet rose above the hills of Namibia. *D. Ewers*

◄ At the end of January 2007, Comet McNaught displayed its splendour in the southern sky. These two panoramic images (left) were taken at the Paranal Observatory of the European Southern Observatory. From one evening to the next, the comet rose higher and higher above the evening twilight and formed an impressive arc in the sky. The Moon is visible at the right edge of the images. *S. Deiries*

Appendix

Comet Hyakutake and an auroral display on 16 April 1996. *Jack Finch*

The greatest comets of modern times

Name	Designation	Perihelion date	Perihelion distance	Distance to Earth	Maximum magnitude	Maximum tail length
Great Comet of 1471	C/1471 Y1	1 Mar 1472	0.49 AU	0.07 AU	−3	50°
Comet Halley 1531	1P/Halley	26 Aug 1531	0.58 AU	0.44 AU	0	15°
Great Comet of 1556	C/1556 D1	12 Mar 1556	0.49 AU	0.08 AU	−2	10°
Great Comet of 1577	C/1577 V1	27 Oct 1577	0.18 AU	0.63 AU	−7	30°
Comet Halley 1607	1P/Halley	27 Oct 1607	0.58 AU	0.25 AU	−1	10°
Great Comet of 1618	C/1618 W1	8 Nov 1618	0.39 AU	0.35 AU	0	100°
Great Comet of 1664	C/1664 W1	4 Dec 1664	1.03 AU	0.17 AU	−1	40°
Comet Kirch 1680	C/1680 V1	18 Dec 1680	0.01 AU	0.49 AU	−10	90°
Comet Halley 1682	1P/Halley	15 Sep 1682	0.58 AU	0.42 AU	0.5	30°
Great Comet of 1744	C/1743 X1	1 Mar 1744	0.22 AU	0.83 AU	−5	90°
Comet Halley 1759	1P/Halley	13 Mar 1759	0.58 AU	0.12 AU	0	47°
Comet Messier 1769	C/1769 P1	8 Oct 1769	0.12 AU	0.32 AU	0	60°
Comet Flaugergues 1811	C/1811 F1	12 Sep 1811	1.04 AU	1.22 AU	0	25°
Comet Halley 1835	1P/Halley	16 Nov 1835	0.59 AU	0.19 AU	1	40°
Great March Comet of 1843	C/1843 D1	27 Feb 1843	0.01 AU	0.84 AU	−10	70°
Comet Donati 1858	C/1858 L1	30 Sep 1858	0.58 AU	0.54 AU	−0.5	40°
Comet Tebbutt 1861	C/1861 J1	12 Jun 1861	0.82 AU	0.13 AU	−3	120°
Great September Comet of 1882	C/1882 R1	17 Sep 1882	0.01 AU	0.98 AU	−17	20°
Great January Comet of 1910	C/1910 A1	17 Jan 1910	0.13 AU	0.83 AU	−5	50°
Comet Halley 1910	1P/Halley	20 Apr 1910	0.59 AU	0.15 AU	−3	240°
Comet Arend-Roland 1956	C/1956 R1	8 Apr 1957	0.32 AU	0.57 AU	−0.5	30°
Comet Ikeya-Seki 1965	C/1965 S1	21 Oct 1965	0.01 AU	0.91 AU	−15	45°
Comet Bennett 1970	C/1969 Y1	20 Mar 1970	0.54 AU	0.69 AU	0	25°
Comet Kohoutek 1973/74	C/1973 E1	28 Dec 1973	0.14 AU	0.81 AU	−3	25°
Comet West 1976	C/1975 V1	25 Feb 1976	0.20 AU	0.79 AU	−3	30°
Comet Halley 1986	1P/Halley	9 Feb 1986	0.59 AU	0.42 AU	2	25°
Comet Shoemaker-Levy 9 1994	D/1993 F2	–	–	–	12	1'
Comet Hyakutake 1996	C/1996 B2	1 May 1996	0.23 AU	0.10 AU	−0.5	100°
Comet Hale-Bopp 1997	C/1995 O1	1 Apr 1997	0.91 AU	1.31 AU	−0.5	25°
Comet McNaught 2007	C/2006 P1	12 Jan 2007	0.17 AU	0.81 AU	−5.5	35°

Glossary

anti-tail   Tail of a comet, projecting in the direction of the Sun, and opposite to the true tail.

aphelion   The point on an orbit that is farthest from the Sun.

astronomical unit (AU)   The mean distance between the Earth and the Sun, corresponding to approximately 149.6 million kilometres.

angular measurement   A means of describing distances on the sky. 360° corresponds to the measurement of the whole celestial sphere, 180° to the portion above the horizon. 1° (degree) corresponds to 60' (arcminutes) or 3600" (arcseconds). The Moon and Sun both subtend approximately 30' (i.e. about half a degree).

circumpolar   Stars surrounding the celestial pole that never rise or set. The radius of this zone is equal to the geographical latitude of the place of observation. In Europe, for example, at a latitude of 50°N, it thus reaches 50° from the North Celestial Pole.

coma   The shell of gas and dust that surrounds the nucleus of a comet.

conjunction   The apparent close approach of two celestial bodies as seen from Earth.

declination   An astronomical co-ordinate, corresponding to latitude on Earth. Its value is 0° at the celestial equator, +90° at the North Celestial Pole and –90° at the South Celestial Pole.

eccentricity   The deviation of a comet's orbit from a circle.

ecliptic   The apparent path of the Sun as projected onto the celestial sphere. The Moon and the planets also move close to the ecliptic.

elongation   The separation between a celestial body and the Sun. Objects with small elongations may be observed only in the daytime sky.

envelopes   Shell-like structures in the region of the nuclei of bright comets. They are created by jets from the rotating nucleus.

ephemerides   Predictions of the positions of celestial objects.

false nucleus   The point of light seen in the head of a comet, as viewed from Earth. It is not the actual nucleus, but the bright innermost region of the coma.

iconography   Research and interpretation of the content and symbolism of pictorial art.

nucleus   The true body of a comet, a mixture of dust and ice, a few kilometres across, and invisible from the Earth.

magnitude   The unit used for apparent brightness. A difference of 5 magnitudes corresponds to an actual difference of 1:100. The faintest stars visible to the naked eye are about magnitude 6, and the brightest star in the sky (Sirius) has a magnitude of –1.4.

meteoroid   A small body in the Solar System, which enters the Earth's atmosphere, and creates a meteor (a 'shooting star').

parallax   The angular difference obtained by the observation of a near object against the distant background (essentially at infinity) when seen from two different locations.

perihelion   The point on an orbit that is closest to the Sun.

quadrant   An instrument for height and angular measurement. It consists of a quarter of a circular plate with a scale divided into angular measurements as well as a plumb line.

right ascension   An astronomical co-ordinate, analogous to longitude on Earth.

striae   Structures in the tail of a comet that arise from repeated release of dust from an active region on the nucleus.

sublimation   The transition from the frozen to the gaseous phase of any substance.

synchrones   Structures in the tail of a comet that arise from the rotation of an active region on the nucleus.

zenith   The point on the celestial sphere directly above the observer.

Bibliography and references

General

- Asimov, I.: *Asimov's Guide to Halley's Comet*, Walker & Co., New York, 1985
- Burnham, R.: *Great Comets*, Cambridge University Press, Cambridge, 2000
- Crovisier, J., Encrenaz, T.: *Comet Science*, Cambridge University Press, Cambridge, 2000
- Eicher, D.: *Comets! Visitors from Deep Space*, Cambridge University Press, New York, 2013
- Griesser, M.: *Die Kometen im Spiegel der Zeiten*, Hallwag, Bern 1985 [in German]
- Holetschek, J.: *Untersuchungen über die Grösse und Helligkeit der Kometen und ihrer Schweife*, Volumes 2–5, Vienna, 1913–1917 [in German]
- Kronk, G., Meyer, M.: *Cometography: A Catalog of Comets* (Volumes 1–5), Cambridge University Press, Cambridge, 1999–2010
- Leitner, B., Pilz, U.: *Kometen, Eine Einführung für Hobby-Astronomen*, Oculum-Verlag, Erlangen, 2013 [in German]
- Levy, D. H.: *Comets: Creators and Destroyers*, Touchstone, New York, 1998
- Levy, D. H.: *Guide to Observing and Discovering Comets*, Cambridge University Press, Cambridge, 2003
- Mobberley, M.: *Hunting and Imaging Comets*, Springer, New York, 2011
- Reichstein, M.: *Kometen: Kosmische Vagabunden*, Harri Deutsch, Frankfurt 1985 [in German]
- Sagan, C, Druyan, A.: *Comet*, Ballantine Books, New York, 1985
- Seargent, D.: *The Greatest Comets in History: Broom Stars and Celestial Scimitars*, Springer, New York, 2009
- Schaaf, F.: *Comet of the Century: From Halley to Hale-Bopp*, Springer, New York, 1997
- Schmude, R.: *Comets and How to Observe Them*, Springer, New York, 2010
- Tamann, G. A., Veron, P.: *Halleys Komet*, Birkhäuser, Basel, 1985 [in German]
- Yeomans, D. K.: Comets: *A Chronological History of Observation, Science, Myth and Folklore*, Wiley, New York, 1991

Internet

- Memorable comets of the past: www.cometography.com/past_comets.html
- Great comets in history: http://ssd.jpl.nasa.gov/?great_comets
- IAU comet resources: http://www.minorplanetcenter.net/iau/lists/CometLists.html
- NASA missions to comets: https://solarsystem.nasa.gov/missions/profile.cfm?Sort=Target&Target=Comets&Era=Present
- Observing comets: http://www.skyandtelescope.com/observing/celestial-objects-to-watch/comets/
- Current and historical comets [in German] http://www.kometarium.com/
- Comets over Nuremberg [in German] http://www.naa.net/ain/personen/kometen.asp

References

Cometary beliefs and fears

- Hamacher, D. W., Norris, R. P.: *Comets in Australian Aboriginal Astronomy*. Journal for Astronomical History and Heritage **14**, 31 (2011)
- Schechter, S.: *Comets, Popular Culture, and the Birth of Modern Cosmology*, Princeton, N.J., 1999

Comets in art

- Olson, R. J. M.: *Fire and Ice: A History of Comets in Art*, Walker & Co., New York, 1985
- Olson, R. J. M., Pasachoff, J. M.: *Fire in the Sky, Comets and Meteors, the Decisive Centuries, in British Art and Science*, Cambridge, 1998

Comets in literature and poetry

- Hartmann, T.: *Cometen Spiegel*, Halle, 1606
- Iffland, A. W.: *Der Komet, Posse in einem Aufzuge*, Leipzig, 1799
- Poe, E. A.: *The Conversation of Eiros and Charmion*, New York, 1839
- Stein, H.: *Der Komet*, Berlin, 2013
- Tolstoy, L., tr. Dunnigan, A.: *War and Peace*, New York, 1968
- Verne, J., tr. Roth, E.: *Off on a Comet*, Philadelphia, 1878
- Wells, H. G.: *In the Days of the Comet*, New York, 1906

Comets in science

- Curtis, H. D.: *The Comet-Seeker Hoax*. Popular Astronomy, **46**, 71 (1938)
- Kippenhahn, R.: *Es began mit Tycho und Halley*. Sterne und Weltraum **3/1997**, 199
- Kracht, R.: *Der Sonne gefährlich nahe, Per Internet Kometen entdecken mit SOHO*. interstellarum Thema **1/2011**, 56
- Leitner, B.: *Kometen entdecken, Sind die Zeiten für Amateure vorbei?* interstellarum Thema **1/2011**, 64
- Meyer, W.: *Das Weltgebäude, Eine gemeinverständliche Himmelskunde*, Leipzig, 1897

Cometary science today
- Brandt, J, Chapman, R.: *Introduction to Comets*, Second Edition, Cambridge University Press, Cambridge, 2004
- Festou, M., Keller, U., Weaver, H. (eds): *Comets II*, University of Arizona Press, Tucson, 2004
- Fischer, D.: *Die Ausschweifenden, Große Kometen und ihre Schweife*, interstellarum **87**, 12 (2013)

Great Comet of 1471
- Jervis, J. L.: *Cometary Theory in Fifteenth-Century Europe*, Dordrecht 1985
- Schöner, J.: *Problemata XVI de cometae (1472) magnitudine longitudineque ac de loco ejus vero*, Nuremberg 1531

Comet Halley 1531
- Apian: *Practica des auff dz 1532 Jar*, Landshut 1532
- Schöner, J.: *Conjectur oder ab nehmliche auslegung Johannis Schöners uber den Cometen so im Augstmonat des MCCCCCXXXI. Jars erschienen ist*, Nuremberg 1531

Great Comet of 1556
- Heller, J.: *Practica, auf das M. D. LVII. Jar, sampt Anzeygung vnnd erclerung, Was die erscheinung, vnnd bewegung, des vergangenen vnnd zuuor angezeygten Cometen Jm sechs und funfftzigsten Jar gewesen, vnd bedeutet habe*, Nuremberg 1556
- Gall, H.: *Bericht über einen Kometen und zwei Erdbeben in Rossanna und Konstantinopel*, Nuremberg 1556

Great Comet of 1577
- Scultetus, B.: *Des grossen und wunderbaren Cometen, so nach der Menschlichen Geburth Jhesu Christi im 1577 Jahr von dem 10. tag Novembris durch den gantzen Decembrem biß in den 13. Ianuary des folgenden Jahrs gantzer 65 tag unter des Monden Sphär uber der Wolcken Region gesehen worden*, Görlitz 1578
- Mästlin, M.: *Ephemeris nova anni 1577, 1576; Observatio et demonstratio cometae aetheri, qui anno 1577 et 1578 … apparuit*, Tübingen 1578

Comet Halley 1607
- Kepler. J.: *Von dem newlich im Monat Septembri und Octobri diß 1607. Jahrs erschienen Haarstern oder Cometen und seinen Bedeutungen*, Halle 1608
- Kepler, J.: *De Cometis Libellis Tres*, Linz 1619

Great Comet of 1618
- Cysatus, J. B.: *Mathematica astronomica de loco, motu, magnitudine et causis cometae*, Ingolstadt 1619
- Kepler, J.: *De Cometis Libellis Tres*, Linz 1619

Great Comet of 1664
- Hevelius, J.: *Podromus Cometicus, quo Historia Cometae anno 1664*, Danzig 1665
- Hevelius, J.: *Descriptio Cometae Anno Aerae Christ. M.DC.LXV. Exorti Cum genuinis Observationibus*, Danzig 1666
- Cassini, G.D.: untit, Philosophical Transactions **1**, 15 (1668)
- Halley, E.: *Astronomiae Cometicae Synopsis*, Philosophical Transactions **24**, 1882 (1704)
- Ray, J.: *Observations Made at Rome, by the Late Reverend Mr. John Ray, of the Comet Which Appeared Anno 1664*, Philosophical Transactions **25**, 2350 (1706)

Comet Kirch 1680
- Bauer, K.: *Regensburg, Kunst-, Kultur- und Alltagsgeschichte*, Regensburg 1997
- Kirch, G.: *Observationes Quaedam Accurate Insignis Cometae Sub Finem Anni 1680 visi, Coburgi Saxoniae a Domino Gottfried Kirch Habitae; Decimo Tertio die Antequam a Quoquam Alio Observatus Sit*, Philosophical Transactions **29**, 170 (1714)
- Kirch, G.: *Neue Himmels-Zeitung, Darinnen sonderlich und ausführlich von den zweyen neuen grossen im 1680 Jahr erschienenen Cometen, deren Gestalt, Grösse, Stand und Bewegung…*Nuremberg 1681
- Halley, E.: *Astronomiae Cometicae Synopsis*, Philosophical Transactions 24, 1882 (1704)
- Dörffel, G. S.: *Neuer Comet-Stern, welcher im November des 1680 sten Jahres erschienen, und zu Plauen im Voigtlande dergestalt observiret worden, sampt dessen kurtzer Beschreibung, und darüber habenden Gedancken*, Plauen 1681
- Dörffel, G. S.: *Astronomische Betrachtung des Grossen Cometen*, Plauen 1681

Comet Halley 1682
- Hevelius, J.: *Extract of a Letter from Mr. Hevelius; being Observations by Him Made at Dantzick, of the Comet which Began There to Appear, Aug.16.1682*, Philosophical Transactions **13**, 16 (1683)
- Halley, E.: *Astronomiae Cometicae Synopsis*, Philosophical Transactions **24**, 1882 (1704)

Great Comet of 1744
- de Chéseaux, P. L.: *Traité de la comète qui a paru en Décembre 1743 & en Janvier, Fevrier & Mars 1744*, Lausanne 1744
- Heinsius, G.: *Beschreibung des im Anfang 1744 erschienenen Cometen, nebst einigen darüber angestellten Beobachtungen*, St. Petersburg 1744

Comet Halley 1759
- Heinsius, G.: *Anzeige daß der im Jahre 1682. erschienene und von Halley nach der Newtonianischen Theorie auf gegenwärtige Zeit Vorherverkündigte Comet wirklich sichtbar sey : und was derselbe in der Folge der Zeit für Erscheinungen haben werde*, Leipzig 1759
- Stoyan, R.: *Atlas of the Messier Objects*, Cambridge 2008
- Rüdiger, G.: *Ein sächsicher Komet*, www.aip.de/~rue/pahlitzsch_new.doc

Comet Messier 1769

- Messier, C.: *Mémoire contenant les observations de la dixième comète observée depius le mois d'aout jusqu'au 1er décembre 1769*, Memoirs de l'Academie Royale 392-444 (1775)
- Messier, C.: *Grande Comète qui a paru a la naissance de Napoléon-le-Grand, découverte et observée pendant quatre mois par M. Messier*, Paris 1808
- Meyer, M.: *Charles Messier, Napoleon, and Comet C/1769 P1*, International Comet Quarterly, **1/2007**, 3
- Cook, J., Green, C.: *Observations Made, by Appointment of the Royal Society, at King George's Island in the South Sea; by Mr. Charles Green, Formerly Assistant at the Royal Observatory at Greenwich, and Lieut. James Cook, of His Majesty's Ship the Endeavour*, Philosophical Transactions **61**, 397 (1771)

Comet Flaugergues 1811

- Herschel, W.: *Observations of a Comet, with Remarks on the Construction of Its Different Parts*, Philosophical Transactions **102**, 115 (1812)
- Schröter, J. H.: *Beobachtungen und Bemerkungen über den grossen Cometen von 1811*, Göttingen 1815

Comet Halley 1835

- Herschel, J.: *Observations of Halley's comet, at the Cape*, Monthly Notices of the Royal Astronomical Society **4**, 25 (1837)
- Schwabe, H. S.: *Der Halleysche Comet*, Astronomische Nachrichten **13**, 145 (1836)
- Smyth, W. H.: *Observations of Halley's Comet*, Memoirs of the Royal Astronomical Society **9**, 229 (1836)

Great March Comet of 1843

- Smyth, C. P.: *Letter accompanying drawings of great comet of 1843*, Monthly Notices of the Royal Astronomical Society **7**, 42 (1846)

Comet Donati 1858

- Donati, G. B., Bruhns, C.: *Osservazioni delle Comete di Donati e di Bruhns, fatte nell' J. R. Osservatorio di Padova*, Astronomische Nachrichten **48**, 357 (1858)
- Bond, G. P.: *On the Figure of the Head of the Comet of Donati, by G. P. Bond Director of the Observatory Harvard College*, Astronomische Nachrichten **56**, 299 (1861)
- Pasachoff, J. M., Olson, R. J. M.: *The Earliest Comet Photographs: Usherwood and Bond for Donati 1858*, Bulletin of the American Astronomical Society **27**, 1331 (1995)
- Gasperini, A., Galli, D., Nenzi, L.: *The worldwide impact of Donati's comet on art and society in the mid-19th century*, Proceedings of the International Astronomical Union **260**, 340 (2011)

Comet Tebbutt 1861

- Schmidt, J. F.: *Beobachtungen des grossen Cometen auf der Sternwarte zu Athen*, Astronomische Nachrichten **55**, 369 (1861)
- Weiß, E.: *Bilderatlas der Sternenwelt*, Esslingen 1882

Great September Comet of 1882

- Tebbutt, J.: *The great comet of 1882*, The Observatory **33**, 252 (1910)

Comet Halley 1910

- Achenhold, F. S.: *Der Halleysche Komet im Jahre 1910*, Leipzig 1985
- Hartmann, J.: *Beobachtungen des Halleyschen Kometen zur Zeit seiner Erdnähe*, Astronomische Nachrichten **184**, 371 (1910)
- Bobrovnikoff, N.: *Halley's Comet in 1910*, Publications of the Astronomical Society of the Pacific **250**, 309 (1931)

Comet Arend-Roland 1956

- Beyer, M.: *Physische Beobachtungen von Kometen. XI.* Astronomische Nachrichten **284**, 241 (1959)

Comet Bennett 1970

- Schmidt-Kaler, T.: *Visuelle Beobachtungen des Kometen Bennett 1969i*, Sterne und Weltraum **9/1970**, 200

Comet Kohoutek 1973/74

- Whipple, F.: *Comet Kohoutek in retrospect*, Proceedings of the American Philosophical Society **120**, 1 (1976)
- Regas, D.: *Remembering Comet Kohoutek*, Sky & Telescope **4/2013**, 32
- Kohoutek, L.: *Die Kohoutek-Kometen, Meine fünf Hamburger Entdeckungen*, interstellarum Thema **1/2011**, 36

Comet Shoemaker-Levy 9 1994

- Stoyan, R.: *Der Tod des Kometen. Amateure dokumentieren den Einsturz von Shoemaker-Levy 9 auf Jupiter und seine Folgen*, Sterne und Weltraum **4/1995**, 312
- Fischer, D., Heuseler, H.: *Der Jupiter-Crash*, Basel 1994

Comet Hyakutake 1996

- Kinoshita, D. et al.: *Ion Tail Disturbance of Comet C/Hyakutake 1996B2 Observed around the Closest Approach to the Earth*, Publications of the Astronomical Society of Japan **48**, L83 (1996)
- Harmon, J. K. et al.: *Radar Detection of the Nucleus and Coma of Comet Hyakutake (C/1996 B2)*, Science **278**, 192 (1997)

Comet Hale-Bopp 1997

- Stoyan, R.: *Vom saturnähnlichen Objekt bis zum Massenselbstmord: Die Gerüchteküche um den Kometen Hale-Bopp im Internet*, Skeptiker **2/1998**

Comet McNaught 2007

- Hönig, S. F.: Striae, *Synchronen und Syndynen, Phänomene im Schweif des Kometen C/2006 P1 (McNaught)*, interstellarum **54**, 34 (2007)
- Leitner, B.: McNaught, *Der Große Komet 2007*, interstellarum **51**, 42 (2007)

Index

Entries in **bold** refer to the main section or box in which the topic is discussed. Entries in *italic* refer to figures.

Aegidius de Lessines, monk: 45

Albertus Magnus (Albert of Cologne): 11

Ammianus Marcellinus: 44

Anaxagoras, philosopher: 10

anti-tail: see tails

Antoniadi, Eugène: 147

Apian (Peter Bienewitz): 27, 34, 51–52, *53*

Apollonius of Myndus: 10

Applewhite, Marshall: 199

Arend, Sylvain: 152–153

Arend-Roland, Comet (1968): **152–55,** *152, 154–55*

Aristotle, philosopher: 10–11, 27, 42, 68

astrological beliefs: **11**, 12, 14, 15, 16, 64

Attila the Hun: 44

Augustus Caesar: *17*

Auzout, Adrien: 74

Aztecs: *11*, 13

Babylonian astronomy: 10–11, 42

Baldung, Hans: 18

Barnard, Edward Emerson: 32, 35, 36, 138, 146, 148

Bayeux tapestry: *17*, 45

Bennett, Comet (1970): **162–65,** *162, 164, 165*

Bennett, John: 163

Bernoulli, Jacob: 81, 91

Bessel, Friedrich Wilhelm: 32, 34, 105, 111, **118,** 122

Biela's Comet (1172, 1826): 31, **33,** 153, 167

Bienewitz, Peter: see Apian

Biermann, Ludwig: 33

Bobrovnikoff, Nicholas: 33, 221

Bode, Johann Elert: 111

Bond, George: 127, *131*

Bopp, Thomas: 197, 198

Borelli, Giovanni Alfonso: 74

Bortle, John: 158

Brahe, Tycho: 27–28, 59, 62

Bredikhin, Fyodor: 32, 134

broadsheets: 13, 15, 51, *57, 60, 61,* 64, *70, 71, 77, 82, 83*

Bruhns, Karl: 127, 221

Burnham, Sherburne Wesley: 146

Calendars: 8, 64, **91**

   Gregorian: 8, 91

   Julian: 8, 91

Cardano, Girolamo: 14, 27

   astrological comet types: 14

Caricatures: 20, *24, 148*

Cassini, César-François: 105

Cassini, Giovanni Domenico: 73

Cassini, Jacques: 97

Catalogues

   Bennett's: 163

   Cape Photographic Durchmust-erung: 139

   Messier's: see Messier, Charles

Cavallini, Pietro: 17

CCD cameras: 36, 178, *182,* 210

Centaurs: 38

Charles V, Holy Roman Emperor: 56

   Charles' Comet: 56, see also Comet of 1556

charts: 5 (see individual comet entries)

   key to: **8**

chemical constituents: 33, 144, 147, 168, 190, 199, 210

Chevalier, João: 102

Chinese astronomy:

   astrology: **12**

   observations: 33, 42, 44, 45–46, 50, 51, 55, 63, 67

Christian interpretations: **13**, 15, 18, 26, 51, 56, 81

   criticism of: 91

Clairault, Alexis Claude: 102

Claudius, Roman emperor: 42

colours (of comets): 74, 134, 153, 163, 172, 190

coma: **39**, 105, 111–112, 134, 153, 190, 198, 218

Comet (unnamed)

   of 905: 45

   of 1106: 29, 35, 45, 80, 138, 157

   of 1264: 45, 56, 127

   of 1402: 45

   of 1471: 18, **49–50,** *49*

   of 1556: 49, **54–57,** *54, 56,* 127

   of 1577: **58–62,** 58, *60–61, 62*

   of 1618: **66–71,** *66, 68, 69–71*

   of 1664: **72–77,** *72, 74, 75–77*

   of 1744: **96–100,** *96, 99–100,* 172, 208, 210

   of 1812: 26

   of Mar. 1843: 35, **120–24,** *120, 122, 123–24*

   of Sep. 1882: 122, **136–41,** *136, 139, 140–41,* 157, 158

   of Jan. 1910: **142–44,** *142, 144,* 147

   of 1947, Southern: 152

   of 1948, Eclipse: 152

'Comet Eggs': **13**, 15

Conrad of Megenberg: 11, 45

conspiracy theories: **16**, 168, 199

Constantinople (Istanbul): 62

   earthquake of 1556: *54,* 56, *57*

Cook, James: 105

Cowell, Philip Herbert: 148

Creti, Donato: *20*

Crommelin, Andrew: 148

Curtis, Heber: 147, 219

Cysat, Johann Baptist: 67–68

d'Angos, Jean Auguste: **31**

da Legnano, Giovanni: see John of Legnano

Danielson, Edward: 178

data: 8, **217** (see also individual comet entries)

de Chéseaux, Philippe-Loys: 97–98, 220

   de Chexeaux's Comet: see Comet of 1744

de Ferrer, José Joaquín: 112

de la Nux, Jean Baptiste François: 102, 103

de Maupertuis, Pierre Louis Moreau: 26

de Munck, Jan: 97–98

de Silva y Figueroa, García: 67

Delisle, Joseph-Nicholas: 102–103

Democritus, philosopher: 10

Desaguliers, John Theophilus: 30

designations: 8, **37**

Diaz, Casimiro: 79

Diodorus, philosopher: 42

'dirty snowball' model: 33, 39

discoverers: 8, 31, **35,** 35 (see also individual comet entries)

   table of: **36**

distance determinations:

   see also parallax

   sublunar vs. superlunar: 50, 62, 64

Donati, Comet (1858): *22, 23, 24,* 31, **125–131,** *125, 127, 128–31*

Donati, Giovanni Batista: 32, 126

Dörffel, Georg Samuel: 29, 80, 89, *90–91*

Dumouchel, Dominique: 117

Dürer, Albrecht: 18, *21*

eccentricity: 8, 218

Ebendorfer, Thomas: 45–46

Edgeworth, Kenneth: 33

Edward VII, of Great Britain: 147

Elkin, William: 138

Encke, Johann: 31, 80

ephemerides: 50, 74, 218

Epigenes, astrologer: 10

Euler, Leonhard: 80, 102

Fabricius, Paul: 55, 56

fear of comets: **10–16, 26,** 42, 45, 81, 91, 127, 147

   linking with natural disasters: **14,** 18, 26, 45, 54

   poisoning of atmosphere: 64, 147

   precursors of death: 11, 13, 14, 52, 62, 147

Finlay, William: 138

Flamsteed, John, 79–80, 90

Flaugergues, Comet (1811): 26, **110–115,** *110, 113–15*

Flaugergues, Honoré: 111

fragmentation (of comets): 32, 35, 67, 121, 138, 158, 167, 172, 184–185

Fraunhofer, telescope: 118, 127

Galileo (Galileo Galilei): 28, 68

Galileo spacecraft: see space probes

Gauss, Carl Friedrich: 30

Gemma, Cornelius: 27, 55, 62

Genghis Khan: 45

German mile (Meile): **46,** 50, 62

Giotto di Bondone: *18,* 45

Giotto spacecraft: see space probes

Gould, Benjamin: 138

Grassi, Orazio: 68

Grosseteste, Robert: 11

Guoy, A.: 139

Hagecius, Thaddaeus (Tadeáš Hájek): 27, 62

Hale, Alan: 198

Hale-Bopp, Comet (1997): *Cover image*, 16, 112, **196–207**, *196*, *200–207*, 210

Halley, Edmond: 29, *30*, 80, 90–91, 102, 118

Halley's Comet: **34**
  of 12 BC: 42, 44
  of AD 66: 42
  of 218: 42
  of 374: 42
  of 451: 42, 44
  of 530: 45
  of 837: 34, 45
  of 1066: *17*, 45
  of 1145: 45
  of 1222: 45
  of 1301: *18*, 45
  of 1378: 29, 45
  of 1456: *18*, *27*, 29, 45
  of 1531: 27, 29, 34, **51–53**, *53*, 91
  of 1607: 29, *29*, 34, **63–65**, 91
  of 1682: 28, 34, 53, **90–95**, *92–95*
  of 1759: 15, *30*, 34, **101–104**, *104*, 117
  of 1835: 34, **116–119**, *116*, *119*
  of 1910: 16, 24, 25, 26, 144, **145–151**, *145*
    149–51, 163, 168, 178
  of 1986: *33*, 34, 41, **176–83**, *176–77*, *180*, *181–83*

Halley-family comets: 38

Harold II, King of England: *17*, *18*, 45

Harriot, Thomas: 63

Hartmann, Thomas: 26

Heaven's Gate, sect: 16, **199**, *199*

Heinsius, Gottfried: 97, 98, 101

Heller, Joachim: 55, 220

Herlitz, David: 15, *65*

Herschel, Caroline: 118

Herschel, John: 118, *119*, 122, 134

Herschel, William: 111–112

Hevelius, Johannes: 28, 73, 74, *75*, 90
  Cometographia: 91

Hind, John Russell: 56, 127

Hippocrates of Chios: 10

Hoffmann, Christian Gotthold: 101

Hollywood films: 26

Hönig, Sebastian: 36, 221

Hooke, Robert: 73, 74, 79, 90–91

Hubble Space Telescope: 186, *187*, 190, 199

Huggins, William: 32

Humboldt, Alexander: 111

Huygens, Christiaan: 73

Hyatutake, Comet (1996): *4*, 49, 54, **188–195**, *188*, *191–95*, *216*

Hyakutake, Yuji: 189

Iffand, August Wilhelm: 26

Ikeya, Kaoru: 157

Ikeya-Seki, Comet (1965): 34, 45, **156–161**, *156*, *158*, *159–161*

Istanbul: see Constantinople

Jacobus Angelus (Jakob Engelhart): 45

Jet Propulsion Lab (JPL): 178

Jewitt, David: 178

John of Legnano (Giovanni da Legnano): 11

Julius Caesar: 17, 42
  Julium Sidus (the Julian Star): 42
  Shakespeare's play: 26

Jupiter, planet: 11, 29, 37, 38, 184–86, *186*, *187*

Kandinsky, Vassily: *25*

Kepler, Johannes: 28, 34, 44, 63–64, 66, 67, 68, 220
  astrological beliefs of: 16, 28, 64

Kepler's Laws: 30, 37, **39**

Kirch, Comet (1680): 13, 15, *20*, 29, 42, **78–89**, *78*, *82–89*, 90, 91, 99

Kirch, Gottfried: 79

Klinkenberg, Dirk: 97
  Klinkenberg's Comet: see Comet of 1744

Kohoutek, Comet (1973–74): 16, 164, **166–69**, *166*, *168*, *169*, 221
  disappointment with: 163, 164, **168**, 172

Kohoutek, Luboš: 32, 167, *168*

Korean astronomy: 45, 50, 67

Kreil, Karl: 118

Kreutz Group (Kreutz Sungrazers): **35**, 121, 138, 157

Kreutz, Heinrich: 35

Kron, Erich: 148

Kronk, Gary W.: 5, 157, 219
  Cometography: 5, 219

Kuiper Belt: 33, 38

Kuiper, Gerard: 33

Lalande, Jerome: 16, 102

Levy, David H.: 36, 179, 184, 185, 219

light pollution: 40, 41, 168

London, Plague of: 74

long-period comets: 37, 38, see also sungrazers

Louis the Pious: 45

Louis XV, King of France: 31, 106

Lovejoy, Comet (2011): *35*

Lowell, Percival: 147

Lubinietzky, Stanislaus: *15*, 28
  Theatrum Cometicum: 28, *29*, 43, *46–47*

Luther, Martin: 15, 51

Machholz, Don: 36

Maclear, Thomas: 122

Mädler: Johann Heinrich: 127

magnitude: 8, **41**, 217, 218

Mann, William: 127

Maraldi, Giovanni Domenico: 105

Marcus Agrippa: 42

Mars, planet: 11, 14, 16, 147

Marsden, Brian: 35

mass-loss: 34, 118

Mästlin, Michael: 27, 59, 220

Maximian of Ravenna, throne of: *17*

McNaught, Comet (2007): 98, 158, **208–215**, *208*, *211–15*

McNaught, Robert: 35, 36, 198, 199, 209, 210

Mercury, planet: 11, 14, 30

Messier, Charles: 30, 31, 34, 36, 98, 101–103, *103*, 105–106
  Messier 1 (The Crab Nebula): 30, 178
  Messier Catalogue: 30, 103, 163, 220

Messier, Comet (1769): **105–109**, *107–109*

meteor showers: 10, 31, 32, 41, 147, 218
  Bielids: 32
  Eta Aquarids: 34, 45
  Orionids: 34

Meyer, Maik: 5, 36

Miller, William: 122

Millerites: 122

Mithridates VI, King of Pontus: 42

Montaigne, Jacques: 32

Morehouse, Comet (1908): 147

Mrkos, Antonín: 36, 153

Müller, Johannes: see Regiomontanus

Myers, David: 168

names: see designations

Napoleon: 20, 106, 112

Narbut, Georgii: *25*

Neptune, planet: 33, 38

Neptune-family comets: 38

Nero, Roman emperor: 42, 112

Nero's Comet: 42

Nettles, Bonnie: 199

Newton, Isaac: 73–74, 80, *89*, 90
  Principia: 29

Nixon, Richard: 168

Nuremberg: 5, *62*, *82*, *83*, *84*, *95*, *107*

nomenclature: see designations

nucleus: 33, **38–39**, 40, 41, 138, 141, 151, *172*, 218
  envelopes: 127, 134, *135*, 138, *141*, 163, 198, *207*, 218
  false nucleus: 41, 207, 218
  jets: 40, 79, 98, 118, 122, 134, 163, 179, 190
  outbursts: 40, 45, 91, 146, 147, 179, 190, 198

O'Meara, Stephen James: 179, 210

Olbers, Heinrich Wilhelm: 30

Oort Cloud: 33, 38, *29*, 143

Oort, Jan: 33

Ophey, Walter: *25*

Öpik, Ernst: 33

orbits: 8, 37
  early orbit determinations: 28–32, 74, 80, *89*, 91
  precise calculations: 30, 102, 105, 118, 148, 184

Palitzsch, Johann Georg: 30, 34, **101**, *103*, 146

Palmer, Samuel: *23*, 24

pamphlets: see broadsheets

Paracelsus, physician: 52

parallax: **27**, 46, 52, 62, 218

perihelion: 8, 28, 29, 30, 32, 35, 37, 40, 216, 218

Petit, Pierre: 74

Petrus Lacepiera (Peter of Limoges): 45

Philostorgius, historian: 44

photography:
  early: 34, 35, 36, 125, 127, 138–139
  telescopic: 127, *130*, *131*, *140*

Pingré, Alexandre Guy: 73, 80, 105

Pizarro, Francisco: 59

Pizarro, Guido: 171

Pliny the Elder: 11, 12, 14, 16

Poe, Edgar Allan: 26

Pons, Jean Louis: 31

Ptolemy, Claudius: 11, 13

   Tetrabiblos: 11, 16

Pythagoras, philosopher: 10

quadrant: 27, **88**, 218

Ray, Peleg: 121

Regiomontanus (Johannes Müller): 27, 50

Reslhuber, A.: 127

Roland, Georges: 152–53

Sarjan, Martiros: *25*

Saturn, planet: 11, 14, 16, 18, *21*, 28, 44, 74, 199

Schiaparelli, Giovanni: 32, 147

Schilling, Diebold: 18

   Lucerne Chronicle: *18*

Schleusinger, Eberhard: 50

Schmidt, Julius: 127, 134, 221

Schröter, Johann Hieronymus: *112*, 221

Schwabe, Heinrich: 127, 221

Schwarzschild, Karl: 148

Scultetus, Bartholomäus: 59, 220

searches, automated: 32, 33

   Spacewatch programme: 33

   table of: **35**

Seki, Tsumoto: 157

Seki-Lines, Comet (1962): 158

Seneca, philosopher: 10, 11, 13, 42

   death of: 42

Seven Year's War: 103

Shaanxi, earthquake of 1556: 56

Shakespeare, William: 11, 26, 42

Shoemaker, Carolyn and Eugene: 184–86

Shoemaker-Levy 9, Comet (1994): **184–187**, *184, 185*

shooting stars: see meteors

short-period comets: 31, 35, 37, 38

Shramek, Chuck: 199

Silbernagel, E.: 143

Skjellerup-Maristany, Comet (1927): 156

Smyth, Charles Piazzi: 122, *123*, 221

Smyth, William: 112, 118, *127*, 127, 221

space probes: 33, **34**, 179

   Deep Impact: 33, 34

   Galileo: **186**

   Giotto: 33, *33*, 34, 41, 179, *180*

   ROSAT: 190

   Rosetta: 33, *33*, 34

   Sakigake: 34, 179

   Skylab (space station): 167–168, *169*

   SOHO: 5, 35, 36, *190*, 210, *211*

   Soyuz-13: 168

   Susei: 34, 179

   Ulysses: 210

   Vega: 33, 34, 179

spectroscopy: 31, 33, 139, 144, 147, 148, 168

speed (of comets): 16, 39, 68, 185

   early measurement: 29, 139

Star of Bethlehem, as a comet: *17, 18, 19, 20, 43*, **44**, 45, 168

Storer, Arthur: 90, 91

Striae: see tails

Struve, Friedrich Wilhelm: 118

superstitions: see fear of comets

sungrazers: 47, 137, 139, see also Kreutz Group

Swift, Lewis: 32, 36

Swift-Tuttle, Comet (1862):

synchrones: see tails

tails:

   anti-tail: 41, 80, 138, 153, *155*, 167, *169, 200*, 210, 218

   dust tail: 40, 41, 127, 143, 146, 153, *165, 174*, 179, 190, 196, 198

   dust trails: 41, 122

   gas (or plasma) tail: 33, 40, 122, 127, 143, 146, 148, 153, 158, 163, *165*, 172, 179, 190, 210

   shapes: 31, 75, *74*, 134, *149*, 158, 210

   striae (and rays): 41, 98, 143, 158, *159*, 172, 198, 208, 210, 218

   synchrones: 40, 172, 198, 218

Takamizawa, Kesao: 189

Tebbutt, Comet (1861): **132–135**, *132, 134, 135*

Tebbutt, John: 132–33, 221

telescopic observations (early): 27–28, 66–67, 74, 79–80, 98

Tempel, Comet (1864): 32

Tempel, Wilhelm: 36

Tempel-Tuttle, Comet (1866): 32

Theophanes, Byzantine chronicler: 44

Theophrastus, philosopher: 42

Thirty Years' War: 15, 66, 68, *68*, 72

Thollon, L.: 139

Thome, Juan: 138

Tolstoy, Leo: 26, 219

Toscanelli, Paolo: 46, 50

Turner, William: *22*, 24

van der Stel, Simon: 91

Venus, planet: 11, 14, 44, 59, 74, *151*

Verne, Jules: 26, 219

Verschuier, Lieve: *20*

Vespasian, Roman emperor: 42

Vienna, siege of: 15, 81, 91, *92*

visual magnitude: see magnitude

von Biela, Wilhelm: 32, see also Biela's Comet

von Boguslawski, Palm Heinrich Ludwig: 118

von Peuerbach, Georg: 27, 45–46

von Zach, Franz Xaver: 111, 112

Wargentin, Pehr Wilhelm: 106

Weigel, Erhard: 81

Wells, H.G.: 26, 219

West, Comet (1976): 98, **170–175**, *170, 173–75*, 198

West, Richard: 171

Whipple, Fred: 33, 39, 40, 127, 166, 168, 221

William the Conqueror: 45

Wolf, Max: 143, 146, 147, *149*

Zwingli, Huldrych: 52

Figure credits